a Consumer Publication

CENTRAL HEATING

Consumers' Association
publishers of **Which?**
14 Buckingham Street
London WC2N 6DS

a Consumer Publication

edited by Edith Rudinger

published by Consumers' Association
publishers of **Which?**

Consumer Publications are
available from Consumers'
Association, and from
booksellers.
Details of other
Consumer Publications
are given at the end of this
book.

© Consumers' Association April 1975
revised edition October 1975
September 1977
August 1979

ISBN 0 85202 137 2

 Computer typeset
and printed offset litho
by Page Bros (Norwich) Ltd

CONTENTS

Warmth and comfort	*page*	1
insulation		4
Choice of fuel		9
cost and availability		16
Wet systems		20
Boilers		22
Radiators and convectors		35
Circulation		41
Controls		47
thermostats		49
valves		53
Domestic hot water		58
Warm air heating		64
Electric heating		71
The installing		76
Problems and difficulties		89
Avoiding hazards		100
Obligations and requirements		104
Index		108

instead of a FOREWORD

'The present trend towards higher indoor temperatures could be checked, or even reversed, if people were persuaded, by fashion or by the cost of heating, to wear more clothing. A reduction of 1°C in the average temperature could result in a 10 per cent saving in fuel usage, and this requires, for instance, no more than a return to the wearing of a waistcoat.'

(from the Billington report*)

Domestic Engineering Services, a report prepared for the Institution of Heating and Ventilating Engineers under the chairmanship of N S Billington, then Director of the Heating and Ventilating Research Association

Heating up a room to a high temperature is not necessarily enough to make you comfortable. But temperature is the factor that is most easily measured, so when you buy central heating, the installer should guarantee that certain temperatures will be reached in the various rooms.

The amount of heat you need for comfort depends on how chilly a person you are and how other members of the household react to the cold: for instance, an old person needs more heat to keep warm than a young one, and a baby's room should be kept at a constant temperature day and night. An active person rushing around the house all day needs less heat than a desk-bound one.

What you use the room for affects the temperature you will want in it. You must discuss with the installer the maximum temperatures that you are likely to need in each room, and the central heating system must be designed to reach them. For instance, in a living room or dining room you should be able to have a temperature of about 20 °C (68 °F); a bedroom generally needs only 16°C (61°F) but if it is used as a study or playroom in the day time, you should be able to warm it up to the same temperature as the living room. A kitchen usually has extra heat from the cooker and other appliances, so will not need more than about 17 °C (63 °F), though a large kitchen that is used as a kind of living room during the day, when no cooking is being done, should have heating equipment to bring it up to living room temperatures.

The bathroom should be warm—perhaps the warmest room in the house, because you are sure to be in it without any clothes on at some time or other; also, warmth in the bathroom counteracts condensation. Even though one does not spend much time in the hall or corridors, the temperature there should reach about 16°C (61 °F). Rooms that are temporarily unused, such as the spare room, should not be allowed to get too cold—not below 10°C (50°F)—otherwise there is a risk of dampness. In an elderly person's house, the temperature should be constant throughout the rooms at not less than about 21°C(70°F).

When you have decided what temperatures you want, your central heating system should be able to provide them irrespective of the temperature outside. The average January temperature in Britain is between about 3°C and 7°C, but it can get colder and central heating installers generally work on the basis of −1 °C (30 °F). If it is warmer outside, the heating does not have to be fully on. The actual temperatures can be selected and set by the appropriate controls. So, a properly designed system will achieve the required room temperatures at an outside temperature of −1 °C. Only when the outside temperature is below this is there a chance that you will not be warm enough. Spells of even very cold weather should not make the house too uncomfortable if the central heating has been going for some time, because the fabric of the house—walls, furniture and so on—will have been warmed up.

Taking a lower base point than −1 °C would be excessive for what is generally needed, and the system would be uneconomical and inefficient. If, on the other hand, a higher base point were taken, there might be many days on which the system could not cope.

Your central heating system (the number of radiators, the kind of boiler and so on) has to be designed on the basis of being able to heat the rooms to the temperatures you want, with −1 °C outside. The temperatures to be reached in each room are called the design temperatures. If the house is particularly exposed, the installer has to take into account the extra heat losses from your house when calculating the equipment you will need.

Full central heating is designed to heat the whole house to fully comfortable temperatures. Anything less than full central heating is by definition going to be less comfortable. It is possible to install partial central heating in which only some of the rooms have radiators. Selective heating means all rooms can be fully heated, but not all at the same time. If you decide to have low design temperatures, you will get only background heating when it is really cold outside, so you will need supplementary heating, from say, a plugged-in fire. Make sure that when you have a central heating system installed, you know what you are getting: if you want full central heating, do not accept anything less.

You can have a central heating system running continuously or designed and controlled so that it is on only at certain hours. If you heat the house intermittently rather than continuously, there will inevitably be warming up and cooling down periods. In a house whose fabric would not be able to store much heat anyway, it matters less if the heat is turned on only when it is required and off immediately it is not. This can, in fact, be useful in a household where everybody goes out in the morning and comes back in the evening. On the other hand, a solid house with thick walls is better heated continuously because the fabric will take a long time to heat up, but then stores heat.

Your heating may reach the design temperatures, which are taken at head level, but if your feet are and remain icy cold, you will not be comfortable. The radiators or other heat emitters must therefore distribute the heat as evenly as possible throughout the room. Some types of heat emitters are better at doing this than others.

Comfort in a room also depends on the warmth of the surrounding surfaces, such as the temperatures of the walls and the windows. Rooms with large areas of glass, even if well-insulated, will therefore need more heating than rooms with very small windows. Heaters placed against outside walls or under windows will counteract cold air coming off them.

Heat created by central heating should be retained in the house as much as possible, but there must be a certain amount of ventilation and when the installer designs a system he takes into account the so-called air changes. Humidity—too much or too little—is generally not a particular problem that affects comfort, but it is worth making sure that air can be extracted from a kitchen, for instance, where the humidity might be higher than the rest of the house.

With some central heating systems, there is nothing in the room to warm your hands by—no fire or hot surface. Although this type of heating can produce as high a temperature as any radiator, some people may not feel as comfortable if they cannot see or touch something hot. It is a good idea to see how you like any type of heating in someone else's house, before installing it in yours.

Insulation

The better a house is insulated, the lower will be the running costs of any heating installation, because more warmth will be retained and less heating needed. This is particularly relevant when fuel prices are high and rising, and resources of fuel are diminishing.

Once a house has been heated to the temperature you want, almost the only function of the central heating system is to keep topping up the heat as it is lost through the walls, roof, windows, floors and doors. Fresh air which comes in to replace the stale air inside the house has to be warmed up, too.

Before installing any central heating, make sure that your house has good insulation, in sufficient quantities and in the right places. What you spend on extra insulation will be recouped by lower heating costs. There are specialist firms who install various forms of insulation, but some you can put in yourself.

roof

In an uninsulated house, about a quarter of the total heat lost goes through the roof. Good loft insulation saves up to three-quarters of that loss; this is where for the least cost you will get the greatest heat saving.

Most roof insulation is easy enough for you to do-it-yourself. The different materials vary in effectiveness. For effective roof insulation with mineral wool, glass fibre, or expanded polystyrene, you need a thickness of 80 mm ($3\frac{1}{8}$ in); more, up to 150 mm (6 inches), for vermiculite. Insulating materials are available in rolls for laying between or over the ceiling joists, or in rigid sheets for laying down or nailing to the underside of sloping roof timbers. Pellets and granules, which you buy in bags, should be spread to an even thickness between the joists (they sometimes drift due to draughts in the roof space). Many houses have a 25mm (1 inch) layer of insulation material—increasing it to 80mm considerably increases the insulation. But adding indefinitely to the thickness does not go on adding economically to the effectiveness of the insulation.

Aluminium foil is also used for insulation. This can be unbacked or bonded to reinforced paper which is sometimes corrugated; alternatively there is foil-backed plasterboard. Aluminium foil can be laid across the joists and tacked down; corrugated foil can also be laid between the joists. To be effective, the shiny side must be facing downwards. A layer of corrugated foil is about as effective as 25 mm of expanded polystyrene (but with foil, you can use only one layer effectively—there is no point in putting down two or three layers).

The whole of the loft floor area should be insulated with the exception of the area directly below any water cisterns, otherwise they might freeze up during prolonged cold spells.

To counteract any risk of condensation, make sure that the roof space is ventilated to the outside.

walls

More heat is lost through the external walls of an uninsulated house than through any other part—probably about 35 per cent of the heat lost. Many houses built since the 1930s have external cavity walls, that is an inner and outer wall (known as leaves) with an air space, usually of 50 mm (2 inches), between them. The layer of air between the leaves provides a certain amount of insulation. This can be increased by filling the space with insulating material. The types of materials in common use for cavity filling are loose pellets or mineral wool, or foamed plastic. They are injected into the cavity through a number of holes drilled through the external leaf of the wall in existing houses (or through the internal leaf while the house is being built). The loose fills are blown in under pressure to pack the space. The plastic is injected in a liquid state, foams up, and solidifies into an aerated rigid substance. Afterwards the holes are filled in with mortar, or if the holes have been drilled with a hollow core drill, the brick cores are replaced.

Wall cavity insulation is fairly expensive but once you have it, it can stay there for the life of the building. The fuel savings can outweigh the initial costs within a few years.

The work has to be done by a contractor specialising in this type of insulation. Different types of fillings are branded; some firms do one, some another. The insulation in its final state cannot be inspected. Some of the foams used to prove somewhat unreliable: they shrank exces-

sively, leaving cracks and gaps which allowed moisture to bridge the cavity. Cracking is less with the low shrinkage foams now available.

There is a risk of dampness spreading from the outside in walls frequently exposed to severe driving rain. (Protecting an exposed wall with a waterproof finish can counteract this.) You can write to the Building Research Advisory Service, Garston, Watford WD2 7JR for technical information on the advantages and possible risks.

Most cavity wall insulation firms now have the Agrément Board certificate. (The Agrément Board was set up by the government to provide independent assessment of building materials, products, components and processes.) With such a certificate, the local authority has merely to be notified prior to the work being carried out. Without it, you would need specific permission from the local authority.

A house that does not have cavity walls is less well insulated to start with. Solid walls can be insulated by adding some form of internal lining, such as expanded polystyrene stuck to the plaster surface; the thickness of the insulation is limited and the improvement is not likely to be spectacular. Thicker lining, such as expanded polyurethane laminated to plasterboard, is likely to improve the insulation more. Less straightforward but more effective is fixing insulating board or aluminium foil-backed plasterboard, or a 'sandwich' of plasterboard with expanded polystyrene in between, is likely to improve the insulation more. Less straightforward but more effective is fixing insulating board to battens fixed to the walls, so as to create an insulating air space. There will, however, be difficulties with decoration, skirting boards and mouldings. Solid wall insulation should include a vapour barrier, such as polythene sheeting, to prevent condensation forming behind it.

windows

Two panes of glass (instead of one) with a layer of still air between them improve insulation. Double glazing as compared with single glazing can save about half the heat loss through the window. But this fifty per cent saving relates just to the windows—through which only about a tenth of the heat of a house may be lost (the amount varies considerably according to the ratio of window to wall).

There are other benefits apart from cutting down heat loss. With double glazing, the inner pane will be at a higher temperature than the outer one, so that cold downdraughts are reduced and the area round the window is warmer, which makes the whole room more comfortable. Also, double glazing can reduce condensation on the window, and some noise transmission from outside.

For heat insulation, the air gap between the panes of glass should preferably be about 20 mm ($\frac{3}{4}$ inch). The best thickness of the layer of air between the panes depends on the size of the window and the temperatures involved. Too narrow a layer of air (less than about 12 mm, $\frac{1}{2}$ inch) lets heat be conducted out through the glass; too wide a gap allows air to circulate. 20 mm would not make much difference to the amount of external noise transmitted into the house. For effective sound-proofing, the gap would have to be at least 100 mm (and up to 200 mm), which is still satisfactory for heat insulation.

There are many different types of double glazing available. Factory-sealed units are two panes of glass permanently sealed together all the way round. The unit replaces a single pane of glass in the frame. This is an unobtrusive and effective means of double glazing. Other methods, for both do-it-yourself and professional installation, are fixing second panes of glass or plastic to the existing window (non-opening), or adding hinged or sliding panels (openable). Double or coupled windows replace a complete window, frame and all. They involve structural alteration, and are therefore best fitted when a house is being built or when existing frames need replacing.

other insulation
The best form of insulation for floors is a good fitted carpet, complete with underlay. This considerably reduces the heat being lost downwards and any cold air coming up. If a carpet is fitted right to the skirting boards, any draughts from under the skirting boards will also be eliminated. If you do not have fitted carpets and there is sufficient room below the floor joists, insulating material can be fitted in the space. There is no need to worry about insulating the floor between rooms which are both heated— for instance a bedroom above a living room.

Windows that open should be a good fit in their frames—sealing strips help to stop draughts. External doors should have a draught excluder, and an enclosed porch is an advantage. A lot of heat is lost through an open fireplace. This can be reduced by having a throat restrictor fitted in the flue (which also helps to improve the efficiency of an open fire).

Heating pipework which does not provide any useful heating surface should be insulated. (This is less important for pipes fitted between floors where any heat loss will be into the house.) Different materials can be used for pipe insulation. For instance, flexible preformed sleeves made from expanded plastic foam can be slipped on to the pipes during or after installation. Rigid insulation, such as preformed mineral wool or glass fibre, is usually held on to the pipes with metal clips fixed at intervals. Materials are also available in rolls to be wound round the piping.

A boiler outside the house—for instance in a garage or outhouse—should be fitted with an insulated metal jacket, if feasible. This is expensive but the loss from an uninsulated boiler can be equal to that from a large radiator.

Where radiators are positioned against uninsulated external walls, it is a good idea to have a piece of radiator foil fitted to the wall behind the heater.

Hot water cylinders should be fitted with insulating jackets and any cistern and pipes in the roof space must be insulated to prevent freezing. Rigid sheets of expanded polystyrene can be held together with clips round the cistern or a glass fibre blanket, cut to size, can be wrapped round the cistern. Never put any insulation underneath the cistern which could shield it from warmth from below.

A report on insulation was published in *Handyman Which?* in August 1975; other relevant reports include loft and wall insulation and draught excluders in August 1978 and d-i-y double glazing in November 1977.

ventilation

When you insulate your house to keep the heat in, you must not keep all the fresh air out. Some fresh air is needed to keep any fire—including that in the boiler—burning. Also, there must be adequate ventilation (not draught) for people to be comfortable inside the house. So do not block up air bricks, for instance, or make all the windows unopenable.

To a certain extent, the fuel you choose will determine which type of system you can have, so it is a major decision. The choice is basically between solid fuel, gas, oil or electricity.

The two main factors to consider when choosing the fuel are cost and convenience; you will have to live with the effects of both throughout the life of your central heating. To choose only on the basis of the initial cost of installing the system could result in years of dissatisfaction and high running costs, or the expense of alterations and conversion of the system after it has been installed.

solid fuel

Solid fuels for central heating are normally the naturally smokeless fuels (anthracite and Welsh dry steam coal), the manufactured smokeless fuels (such as Coalite, Rexco, and Sunbrite), and coal. All these fuels may be used in a smoke control area, but coal must be burned in appliances approved for the purpose.

Makers of boilers, of roomheaters, and of open fires used for central heating generally list the types and sizes of fuel suitable for their appliances. An illustrated guide to these fuels and their use, *The Right Choice*, is available free from the Solid Fuel Advisory Service, either at the head office Hobart House, Grosvenor Place, London SW1X 7AE or at any local SFAS office.

If you want a central heating system that you switch on and then ignore, solid fuel is not for you. The boiler has to be stoked once a day or more—the number of times depends on the type of boiler and the amount of heat needed in the house. The ash or clinker has also to be cleared away by hand, and disposed of.

—storage

Solid fuel should be stored either indoors or in a covered bunker. If you stoke a boiler with damp fuel, useful heat will be wasted by the boiler having to drive off moisture in the form of steam. Also, solid fuel when wet tends to corrode most metals. This is likely to harm the boiler, and any metal container used for handling or storing the wet fuel will be gradually eaten away.

Solid fuel needs a rather large bunker for storage, which would be a problem for people with a small garden and rules out many flats. The larger you can have the bunker, the better—ideally it should house at least 1½ tons. Some merchants charge less for large deliveries, usually a ton or over.

Equipped with a well-filled fuel store, you can be independent for long periods, which can be important in remote areas in a severe winter—and can also be important when strikes threaten fuel supplies.

Solid fuels vary in density: anthracite is the most dense and gives out most heat for its volume, and coke the least. So if coke is being used, 1½ tons would need a 3·2 cu m (112 cu ft) capacity bunker while 1½ tons of anthracite would need 2 cu m (70 cu ft). If 1½ tons of coke is used up in eight weeks in a central heating system, 1½ tons of anthracite would last about ten weeks in the same system.

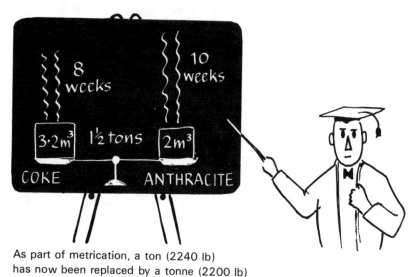

As part of metrication, a ton (2240 lb)
has now been replaced by a tonne (2200 lb)

1 tonne = 1000 kg

For your convenience, the fuel should be stored near the boiler. Deliveries are in sacks which have to be carried, so the bunker should not be too far from the road—preferably not more than 50 metres from where the delivery lorry can unload.

Bunkers should be adequately ventilated to allow damp fuel to dry out and to prevent condensation. The filling hatch at the top of the bunker has to be big enough for the whole of the mouth of a coal sack, and should be positioned where the fuel can be tipped easily, at a height of about 1·4 metres (4 feet 6 inches).

gas

Practically all gas central heating systems are run on piped gas supplied by the gas regions. (But gas is not available everywhere.) Standing charges and tariffs vary in different areas. Ask you local gas region what its tariffs are. The unit cost generally works out less the more gas you use, because the first block of therms each quarter is charged at a higher rate than the final therms.

Natural gas is non-toxic but just as explosive as the previously used town or manufactured gas. It is more or less odourless in its raw state, but is given a smell so that leaks can be more easily detected.

Gas burns without any smoke and with very little smell from the products of combustion. The gas burner should be cleaned and the setting checked at least once a year by a competent fitter to make sure it burns efficiently.

There is no storage involved with piped natural gas. The gas region provides and fixes the service pipe from the gas main in the roadway to the meter in the house, and also the meter itself. They remain their property and responsibility. Where a very long service pipe is necessary, you may be asked to pay a charge towards the installation costs. Where gas is to be brought to a house for the first time, the gas region will want to fix the meter as near as possible to where the service pipe enters the building in order to keep the pipe short. Pipes from the meter to the various gas appliances are the responsibility of the houseowner.

liquefied petroleum gas

Central heating can be run on LPG, but in practice very seldom is. The amount of fuel that would be needed makes even the largest cylinder (47 kg) impractical—and very expensive. Propane can be supplied more cheaply in bulk, to be stored in a tank outside the house, but is still more expensive for running central heating than other fuels—twice the price of piped gas.

oil

The two oils used for domestic central heating are the light 28-second viscosity grade kerosene and the slightly heavier 35-second viscosity grade gas oil. The viscosity number relates to a test where a given quantity of liquid is timed to run through a certain size hole at a fixed temperature. A thicker oil takes longer than a lighter or less viscous oil. The correct grade for a particular boiler depends on the type of burner it has.

Provided the boiler is regularly serviced and there are no leaks of oil or fumes, any smells with oil-fired central heating should be negligible. Regular servicing is necessary to maintain the boiler's safety, reliability and efficiency.

Various suppliers of oil make special offers to anyone willing to sign a delivery contract for a certain period of time, usually not less than one year. Such an offer may include maintenance and servicing at special prices, or subsidised insurance of equipment. Compare what different suppliers have to offer and read the terms of contract carefully before signing anything that ties you to one supplier for a long period of time.

The larger suppliers run planned or automatic delivery services, based on their calculation of the amount of oil you are going to use.

—storage

An oil storage tank should be large enough for at least three weeks' supply—preferably more—at peak winter consumption rate. Its capacity should, if possible, be not less than 1,250 litres (about 275 gallons). If it is large enough to have fuel delivered 2,250 litres (about 500 gallons) at a time, the fuel might be supplied at a cheaper rate.

Delivery tankers can weigh up to some 20 tons, so some drives or roadways are not suitable. The tank must be sited where it can be filled by a hose from the delivery tanker; a terraced property or a house built on a main road where waiting is prohibited can sometimes present difficulties. You can buy ready-made tanks which are usually rectangular or horizontal cylindrical, but you can have one built to a special shape to fit into an awkward space, at extra cost. The bottom of the tank should, if possible, be above the level at which the oil is to be burned, so that it can flow by gravity rather than have to be pumped. It should be sited in the open air, if possible. If an internal tank is essential, it should be in a tank chamber and vented to the outside. Some local authorities require that a large tank be surrounded by a catchpit which can hold the full quantity of oil, plus 10 per cent, in case the tank develops a leak. The relevant British Standard stipulates a catchpit for a storage tank holding more than 1,250 litres, unless the bulk of the tank is buried in the ground or unless the absence of a catchpit does not constitute a hazard.

To protect an oil tank from rust, paint it regularly with an oil-based paint.

The tank should be supported by piers or cradles so that it slopes slightly towards the drain valve at the bottom of the tank. Any sludge or accumulated water which collects at the bottom can be drawn off from time to time. A screwed plug on the valve prevents oil being drained off inadvertently, by children perhaps. The tank must have a vent pipe to let air escape while it is being filled and let air into the tank as the oil is slowly used up during the heating season. The end of the vent pipe should be in the open air where any vapour discharge does not matter. It should be turned downwards so that rain cannot get into the tank.

The vent pipe should be visible from where the tank is being filled, because the first sign of overfilling is oil discharging from the vent. Alternatively, an overfilling alarm system should be fitted that goes off when the oil reaches a certain level in the tank. Some oil suppliers use automatic delivery control which shuts off the flow of oil when the tank is virtually full.

While the tank is being filled—and for about half an hour afterwards—the boiler should be turned off. Otherwise, dirt and impurities (including water) can be stirred up from the bottom of the tank and can get into the burner, causing combustion problems.

A contents gauge or oil level indicator graduated in litres or gallons is more useful than one which tells you only what proportion (such as $\frac{1}{4}$, $\frac{1}{2}$, $\frac{3}{4}$) of the tank is full. There are various types of gauge available, some complex and expensive. A simple one is a vertical, clear plastic tube fitted to a valve at the bottom of the tank, which gives a direct visual indication of the oil level in the tank. The valve stops all the oil leaking out if anything should happen to the plastic tube—only a tubeful would be lost.

The oil supply line from the tank to the boiler must have an oil filter which should be cleaned periodically. There must also be an automatic fire valve and isolating valves close to the tank and to the boiler, to cut off the oil flow in the event of a leakage or a fire, and for servicing.

—centralised fuel storage
On some housing estates where individual tanks would be difficult or impossible to fit, there is a bulk oil tank. The tank is kept full by the fuel company and oil is piped (and metered) directly to the individual users. The consumer does not have to worry about fuel ordering, delivery or storage, but the cost of the oil may turn out not to be competitive.

electricity
There are basically two types of tariff: standard and off-peak. With the standard (full price) tariff, you pay the full amount per unit regardless of when any are used. The standing charge may be distributed over the cost of the first block of units consumed, or added as a lump sum to the quarterly bill.

The alternative to standard tariff is Economy 7 or white meter, with day rate and night rate. The day rate is generally a little more expensive than the standard tariff, but the night rate is cheaper, with Economy 7 being cheaper than white meter. There is generally a higher standing charge than for the standard tariff. White meter rate applies usually between the hours of 11pm and 7am, Economy 7 for seven hours

THE FUELS IN USE

fuel	advantages	disadvantages
solid fuel	can be stored in large enough amounts to make you self-sufficient for long periods	needs daily attention; fuel and ash have to be handled
	possible to get discount on large deliveries	less easy to control automatically than other fuels, some heat output when not required
	price known on ordering, no retrospective fuel bill	needs storage space
		sometimes delivery delays
gas	no fuel shortage or delivery problems	not available everywhere
	can be easily switched on and off automatically	explosive (but has safety controls)
		involves regular servicing
		no choice of supplier
oil	can be easily controlled automatically	fuel must be delivered periodically according to size of storage tank
	can be stored in enough capacity to make you self-sufficient for long periods	storage tank needs careful siting
	price known on ordering, no retrospective fuel bill	involves regular, more complicated servicing than other fuels
		can be smelly
electricity	no fuel storage or ordering problems	no choice of supplier
	no flue or chimney required	supply sometimes cut off
	needs no maintenance	high running costs

between midnight and 8am. New consumers wanting to use off-peak electricity can no longer choose white meter, but anyone can switch to Economy 7.

Another type of off-peak tariff is for electricity used during night hours (with standard rate during the day). This is still in use but no longer being offered to new consumers. A time switch is fitted into the circuit to ensure that off-peak electricity is consumed only during the off-peak times and by heat storage appliances wired into that circuit. In houses which had off-peak electricity before 1970, there may be an afternoon boost available.

With modern off-peak tariff, you can run all appliances all the time but pay a different price per unit according to when you use the electricity. Because of the higher standing charge and higher day rate, it is worth while changing from standard tariff to Economy 7 only if you can use a lot of electricity during the night hours.

Various types of heating equipment have been designed to store the heat produced from electricity which is consumed during the night hours, and to emit the heat during the day. Heat may therefore need topping up towards evening, and this might be more suitable for a household where heat is wanted in the morning and afternoon rather than one where most of the warmth is needed in the evening.

The actual cost per unit within the three types of tariffs varies in different areas. Before deciding on electric heating, check the prevailing tariffs with your local electricity board. Also make sure that the supply cables are of sufficient capacity to carry the load. If they are not, you could be asked for a contribution towards the cost of new ones.

Practically all central heating, irrespective of its heating fuel, is to a certain extent dependent on electricity to operate some parts, such as pumps and fans.

Cost and availability

The table shows the weekly winter running costs of a central heating system giving 525 kWh (about 18 therms) a week heat and domestic hot water—the sort of amount a 3 bedroom semi-detached house might need. The useful heat given out by each fuel depends to a large extent on

RELATIVE FUEL COSTS

	SOLID FUEL						GAS		OIL		ELECTRICITY			
	hand-fired coal (House-warm)		hand-fired coke (Sunbrite)		gravity-feed anthracite grains		final therms		[1]		Economy 7 night rate		Standard	
if you pay....	£55/ton		£66/ton		£66/ton		16.5p/therm		12p/litre		1.14p/kWh		3.07p/kWh	
the cost/useful kilowatt hour will be...	0.95p		1.25p		0.95p		1.01p		1.54p		1.14p		3.07p	
so your weekly winter running cost might be ... (rounded to nearest 5p)	£5.05		£6.55		4.95		£4.10		£8.10		£6.00		£16.10	
	if you pay	cost might be	if you pay	cost might be	if you pay	cost might be	if you pay	cost might be	if you pay	cost might be	if you pay	cost might be	if you pay	cost might be
	£	£	£	£	£	£	p	£	p	£	p	£	p	£
	42	3.85	57	5.65	60	4.50	17.0	4.20	12.5	8.45	1.17	6.15	2.95	15.50
	45	4.10	60	5.95	63	4.75	17.5	4.35	13.0	8.75	1.20	6.30	3.00	15.75
	48	4.40	63	6.25	69	5.20	18.0	4.45	13.5	9.10	1.23	6.45	3.05	16.00
	51	4.65	69	6.85	72	5.40	18.5	4.60	14.0	9.45	1.26	6.60	3.10	16.27
	54	4.95	72	7.15	75	5.65	19.0	4.70	14.5	9.80	1.29	6.75	3.15	16.55
	57	5.20	75	7.45	78	5.85	19.5	4.85	15.0	10.10	1.32	6.95	3.20	16.80
	60	5.50	78	7.75	81	6.10	20.0	4.95	15.5	10.45	1.35	7.10	3.25	17.05
	63	5.75	81	8.05	84	6.30	20.5	5.10	16.0	10.80	1.38	7.25	3.30	17.33
	66	6.05	84	8.35	87	6.55	21.0	5.20	16.5	11.15	1.41	7.40	3.35	17.60
	69	6.30	87	8.65	90	6.75	21.5	5.35	17.0	11.45	1.44	7.55	3.40	17.85
	72	6.60	90	8.95	93	7.00	22.0	5.45	17.5	11.80	1.47	7.70	3.45	18.10
	75	6.85	93	9.25	96	7.20	22.5	5.60	18.0	12.15	1.50	7.90	3.50	18.37
	78	7.15	96	9.55	99	7.45	23.0	5.70	18.5	12.50	1.53	8.05	3.55	18.65
	81	7.40	99	9.85					19.0	12.80	1.56	8.20	3.60	18.90
									19.5	13.15	1.59	8.35		
									20.0	13.50	1.62	8.50		
											1.65	8.65		

[1] average figures used for kerosene and gas oil

the efficiency of the boiler: the table is based on the assumption that solid fuel hand-fired boilers using coke are 70 per cent efficient, those using coal (Housewarm) are 70 per cent efficient, electricity 100 per cent efficient and all others 75 per cent efficient.

Standing charges for gas and electricity vary from area to area; they are not included in the calculations in the table.

Your running costs are likely to be very different from those in the table. They will vary according to the size of house and how much insulation it has; the number of hours that the system runs for each day; the boiler efficiency—which may be less than the one used for the table; the sort of controls the system has; the amount of domestic hot water used—and where you live (temperatures and also fuel prices vary around the country). All these factors have to be taken into account when working out your actual running costs. However, the table can be used as it stands to assess the comparative running costs of the fuels—since the factors relevant to you will affect them all.

Current prices are not the only consideration: you must think about future fuel prices and availability of the different fuels.

In recent years, there have been substantial increases in gas prices. Most of our supplies of natural gas now come from gas fields in the North Sea. Natural gas supplies are finite, but are assured into the next century. Substitutes for natural gas (SNG) have been manufactured to be available for peak demands. SNG also means that we will not all have to be reconverted to town gas if natural gas supplies run out.

About half of our oil is imported from the Middle East. The political situation makes supplies potentially unstable, and prices have continued to go up.

By the end of 1979, three quarters of our oil supplies will be from the North Sea; by the end of 1980 we can hope to be self-sufficient in net terms. A number of new oilfields are being developed; the rise in oil prices makes it economic to develop fields that were hitherto considered unprofitable. Prices are unlikely to go down.

Coal prices depend heavily on transporting costs and wages, neither of which are likely to drop. At present rates of use, we should have enough coal, including house coal and anthracite, for the next 300 years.

Electricity prices are largely dependent on oil and coal—so as their prices go up, so will the costs of electricity. As more nuclear power is introduced, electricity will be less dependent on other fuels; by the end of the century, it is possible that a substantial proportion of the electricity generated will be from nuclear power stations. This will not necessarily make it cheaper.

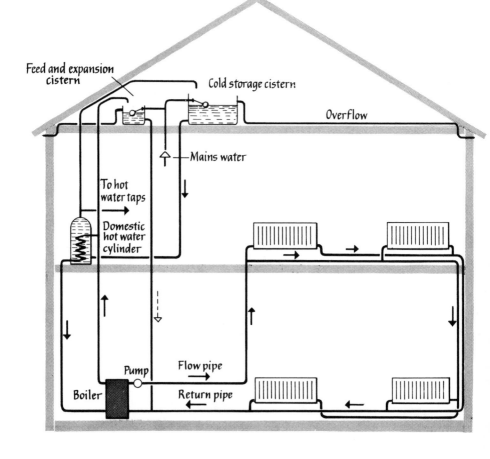

Feed and expansion cistern

Cold storage cistern

Overflow

Mains water

To hot water taps

Domestic hot water cylinder

Pump

Flow pipe

Boiler

Return pipe

In a typical hot water system the heat is provided by a boiler which heats water. The hot water is circulated by a pump through pipes to radiators which give out heat. The cooled water from the radiators is heated up again when it returns to the boiler. The same water is continually going round and round. Any small quantity which evaporates is topped up from a feed and expansion cistern. A pipe from the boiler to the feed and expansion cistern allows the water in the central heating system to expand or contract, and if the water should boil for any reason, the pipe lets the steam escape.

Some of the water that is heated in the boiler circulates (by gravity or, in some systems by being pumped) to the domestic hot water cylinder from which hot water for the taps is drawn. The water which the boiler heats does not go direct to your taps; it passes through a heat exchanger inside the water cylinder and warms up the water in that cylinder. (A heat exchanger is anything which transmits heat from one medium to another, without the two media coming into contact.)

This typical basic system can have many variations and extensions. There are different types of pipework; apart from radiators, the heat emitters can be convectors of different sorts or skirting heating; and there are a number of ways to control the amount of heat for each room and the time when the heat is given out.

In a boiler, fuel is burned in a combustion chamber. The water which is being heated is inside a water jacket—the space between the combustion chamber and the boiler casing—or in hollow metal columns. A flue pipe leads from the boiler to the outside air.

Cast iron boilers, if properly used and maintained, can last 20 years or more. Steel boilers are usually slightly cheaper than cast iron ones but may have a somewhat shorter working life because they are more susceptible to internal corrosion when operated at low temperatures. They are usually enclosed in a stove enamelled, sheet metal jacket which provides some insulation.

Apart from free-standing boilers, there are also wall-mounted and back boilers.

wall-mounted

Wall-mounted boilers are light in weight and very compact because they are designed to contain very little water. They can be fixed or hung on any suitable wall; this is particularly useful where there is not much floor space. Recent types are designed to be fitted within the thickness of an external wall and can be serviced from outside the house, through an access panel.

Wall-mounted boilers are not designed for solid fuel.

With such a boiler, it is imperative that the pump, thermostats and safety controls are operating properly, because the small amount of water can boil very rapidly if any of them should fail.

back boilers

A basic back boiler is a small steel or cast iron box containing water, which fits behind an open fire or is part of a room heater, usually with glass-fronted doors for refuelling. Heat is given directly to the room by a fire or heater, and from the boiler to a pipework system that leads to radiators around the house. Most traditional fireplaces are large enough to take a back boiler, and with the older type of house this was often the only means of heating water. Back boiler systems were originally heated by an open coal or coke fire.

Open fire back boilers, some with a fan-assisted draught, are being re-introduced. There are also room heaters with back boilers designed to run on gas or oil. There must be sufficient ventilation in a room with a back boiler for the fuel to burn properly—more than is needed for a simple open fire.

Heat outputs are somewhat restricted because the size of the boiler is limited by the size of the fireplace recess. A back boiler would probably not be suitable for full central heating in a large house.

size and efficiency

Boilers are described in sales literature not by their dimensions but by their heat output. Heat output is the amount of useful heat extracted from the fuel, which depends on the efficiency of the burner and on the rest of the boiler. For instance, a 17·5 kW (60,000 Btu/h) boiler is capable of producing 17·5 kW (60,000 Btu per hour) from the fuel which it burns. British thermal units (Btu) and therms are measurements of heat output; 100,000 Btu equals one therm.

One Btu is the amount of heat needed to raise the temperature of one pound of water by one degree Fahrenheit. Conversely, it is the amount of heat given out when one pound of water cools by one degree Fahrenheit. The rate at which heat is given out is measured in Btu/hour.

Now, with metrication, heat output is given in watts and kilowatts (kW). 1 kilowatt equals 3,412 Btu per hour. A kilowatt hour is the amount of energy given out in an hour. It has traditionally been used to measure electricity consumption.

Heat input is the amount of heat contained in the fuel which is being burned. The thermal efficiency of a boiler is expressed by its output as a percentage of its input. For instance, a boiler which is 75 per cent efficient gives out as useful heat 75 per cent of the heat contained in the fuel which it is burning.

To work out the size of the boiler you need, the designer has to take into consideration the type of boiler, the type and number of heat emitters, the size of the system, the kind of control system that you are going to have, the amount of insulation, the size of the house and the length of piping (from which inevitably a small amount of heat is lost). It is generally not necessary to add any reserve for heating the domestic hot water except where—perhaps in a family with a lot of children—hot water is going to be used constantly.

There might be a temptation to install a very large boiler with the idea of being prepared for weather that is colder than the design base point of $-1\,^\circ$C. But it is uneconomic to install an oversized boiler which may operate at its full output only once or twice a year and will constantly cycle on and off even during very cold weather. Also, boilers are inefficient when they are run below their full capacity, so the problem is one of boiler efficiency versus the risk of being cold.

It is no use getting a larger size boiler than necessary without rethinking all the other equipment, particularly as to the number and size of radiators.

siting

A free-standing boiler cannot be put just anywhere. The Building Regulations, economy and common sense all affect the siting. Some types of oil-fired boilers are fairly noisy and so should not be put near bedrooms or anywhere where noise could be a problem.

The floor must be strong enough to take the weight of the boiler. Fixing it on a low plinth keeps it away from any dampness when the floor is washed. To avoid fire risk, a solid fuel boiler must not be stood directly on a timber floor and the flue pipe must not be in contact with any woodwork.

There must be enough fresh air to allow adequate combustion. Windows or other openings which could be closed are not enough—there must be permanent fresh air access.

The more centrally in a house the boiler can be placed, the less will be the installation costs because less pipework will be needed. Wherever you put the boiler, it gives out a certain amount of heat. This can be an advantage in, say, a utility room or an otherwise unheated room or a corridor.

If the boiler is also going to heat the domestic hot water supply, it should, if possible, be positioned near the hot water cylinder. To permit good circulation by gravity, the boiler should ideally be directly below the cylinder and the horizontal distance between the two should not be greater than the vertical distance.

All boilers require servicing, so do not have one fitted in an inaccessible place, making servicing difficult or impossible.

flues

Any existing chimney or flue must be large enough to take the products of combustion (flue gases, fumes or smoke) and must be in reasonable condition. The same flue must not be used for more than one appliance burning different fuels—such as an open fire and a boiler: in some circumstances a spark from the coal could ignite gas or oil vapour. What is required by way of lining is controlled by Building Regulations and depends on the construction and state of repair of the existing chimney and the type of boiler. Where a gas-fired boiler is connected to a brick chimney, a suitable liner is normally stainless steel. With an oil-fired boiler, it is useful to have a brick chimney lined with a flexible metallic liner to prevent deterioration of the brickwork. The current practice for solid fuel appliances is to leave an existing chimney unlined; where the house is being built, a clay or concrete liner is built into the chimney.

There are factory-made insulated chimneys which can be installed internally or externally where there is no chimney, or an unsuitable one.

A practical point: before a new appliance or a liner is put in, make sure you have the chimney swept.

—balanced flues

A balanced flue consists basically of a short horizontal metal duct which links the boiler with the outside air. The duct is made in two parts: through one, the air is drawn in for combustion and the other lets out the products of combustion. It is called a balanced flue because whatever the external wind pressure is on the flue terminal, the inlet and outlet have the same, or balanced, pressure.

A balanced flue boiler can often be used where an existing chimney is in very poor condition or in the wrong place or where there is no chimney at all. You cannot fit a balanced flue to an existing boiler—the boiler has to be designed as a balanced flue boiler. Some wall-mounted ones have a fan which assists the air circulation, so that the flue can be smaller. Most balanced flue boilers are designed to run on gas, but some oil-fired ones are available.

A boiler with a balanced flue must be placed alongside an external wall and the terminal on the outside should not be obstructed in any way. The terminal can get very hot so it should be guarded from children, if necessary. Unpleasant fumes are emitted through it, just like through a chimney, so do not have it next to the back door or kitchen window.

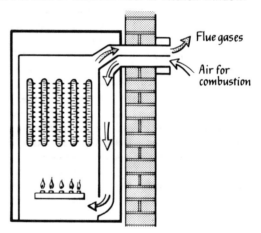

Flue gases

Air for combustion

Even though it takes its combustion air direct from the outside, a balanced flue boiler should not be in a completely unventilated place: there would be a risk of overheating.

solid fuel boilers

Because a solid fuel boiler needs daily attention, it should be put in a place convenient for stoking and removing ash or clinker. This could be the kitchen or a utility room or outhouse.

With a hand-fired boiler, you have to shovel the fuel through the refuelling door on to the fire, or tip it in from a hod whenever more fuel is needed. It is quite a chore. Hand-fired boilers (also called sectional

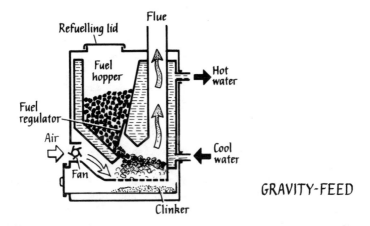

GRAVITY-FEED

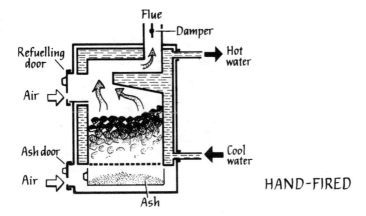

HAND-FIRED

boilers) need refuelling at least twice a day, usually three or four times when it is very cold.

With a gravity-feed boiler, you tip the fuel in at the top and it gradually feeds down through a hopper on to the fire. With this type of boiler it is important to use the correct type and grade of fuel, as specified by the boiler manufacturer. These boilers need refuelling about once a day in winter, but it obviously depends on how cold it is outside and how warm you want the house to be.

Gravity-feed boilers produce mostly clinker, or fused ash. Hand-fired boilers produce mainly ash. Taking hot ashes away is more difficult and unpleasant than removing clinker.

Solid fuel boilers have very few moving parts. They are therefore quiet and there are few things to go wrong. A solid fuel boiler cannot ignite itself or turn itself off automatically or stay alight for any length of time without attention. You cannot for instance, go away for the weekend without leaving a boiler-sitter if you want to come back to a warm house.

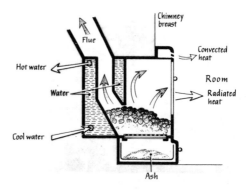

—solid fuel back boilers

Most solid fuel room heaters with a back boiler are hand-fired, but some incorporate a hopper to feed in the fuel. They have glass-fronted refuelling doors through which you can see the fire and through which heat is radiated out. The rate of burning can be controlled manually or thermostatically. Some models have a fan and time switch, so that the heat output can be regulated for when it is most needed. However, it can be difficult to get the right balance between the amount of heat going into the room and the amount being used to heat water for circulation to the rest of the system. Some models have a manually operated damper to adjust this—but only to a limited extent.

Room heaters are designed to burn smokeless fuel but there is an openable room heater in which house coal (which is cheaper than smokeless fuel) can be used smokelessly.

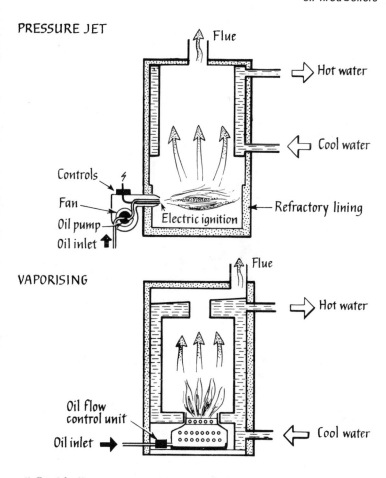

PRESSURE JET

Flue

Hot water

Cool water

Controls

Fan

Oil pump

Oil inlet

Electric ignition

Refractory lining

VAPORISING

Flue

Hot water

Oil flow
control unit

Oil inlet

Cool water

oil-fired boilers

Before oil will burn it has to be atomised into very fine droplets or vaporised. Boilers are designed to do one or the other.

In a pressure jet boiler, the oil is forced through a nozzle by a pump which is part of the burner, so that it comes out as a very fine spray. The oil spray is mixed with air, which is blown in by a fan and burns when ignited. Most pressure jet burners use 28-seconds oil; some can use either 35-seconds or 28-seconds oil.

The various types of vaporising boiler are based on the principle of heating up the fuel so that it vaporises. The burner flame then ignites the fuel which burns as a vapour mixed with air. The simplest type of vaporising boiler is one fitted with a natural draught pot burner. A more sophisticated type is a wallflame boiler. Vaporising boilers should be run on 28-seconds oil.

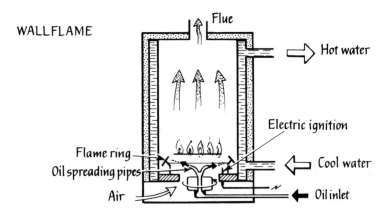

With pressure jet burners, the fan and oil pump are electrically driven by a motor, and with a wallflame there is an electrically driven fan. Natural draught pot burners have no moving parts.

The efficiency of an appliance can be adjusted by varying the amount of oil and air delivered to the burner. Incorrect proportions of oil and air can lead to poor combustion (and soot up the flueways). It is therefore important that the boiler is set up correctly by the installer.

Pressure jet boilers used to be rather noisy and were better sited in an outhouse where this would not be troublesome. Many modern pressure jet boilers are especially designed to be quiet and unobtrusive enough for use in the kitchen. Vaporising boilers are fairly quiet and can be put in a kitchen. Only natural draught pot boilers are completely silent.

Most oil-fired boilers require a conventional chimney or flue but there are some with a type of balanced flue.

By fitting the right controls, an oil burner can be used on a fully automatic basis. But some natural draught pot burners have to be re-lit by hand and cannot be operated fully automatically. Oil boilers can be bought complete with pump, time switch and all essential controls and thermostats.

DOBETA Ltd (the oil burning equipment testing association) tests boilers and other equipment for safety, reliability and performance. Those up to DOBETA's requirements can carry the seal of approval. Make sure that the boiler you get has it.

—oil-fired back boilers

Traditionally, oil-fired boilers have been free-standing, but oil-fired room heaters with a back boiler for fixing in the fireplace are now available. They have a vaporising type of burner, are silent in operation and, if working properly, do not give out any smell of oil. Some are glass-fronted; some have ceramic elements which glow when they are hot.

With some types you cannot regulate the heat between room and water.

gas-fired boilers

A gas-fired boiler can be free-standing or wall-mounted, with either a balanced or conventional (open) flue. Some have a permanent pilot light, others have automatic, electric, ignition to light the pilot. When the boiler starts operating, the pilot lights the main burners which heat up the water in the boiler.

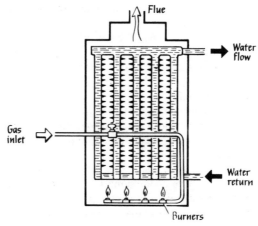

You can buy a gas boiler complete with pump, time switch and controls; alternatively, you can buy a basic gas boiler and have a pump and accessories fitted separately. Gas boilers are quiet and do not smell and are suitable for a kitchen, particularly the small wall-mounted type which does not take up valuable floor space. Some models have a forced draught burner which incorporates a fan and produces a more intensive output for its size than the more usual atmospheric burner. The heat output from a wall-mounted boiler is sufficient for the average size house.

One type of gas-fired system by Servowarm uses a master radiator which is in effect a balanced flue gas boiler, looking somewhat like a very bulky radiator. This has to be placed against or abutting an outside wall. Hot water is pumped from it to conventional radiators in other parts of the house. But before you have it installed, find out about servicing arrangements; gas regions would not service it if anything went wrong.

—gas-fired back boilers

Unlike the other types of back boiler, a gas-fired back boiler and room heater are separate and can be controlled independently. The gas fire gives out enough heat for an average living room. It need not be fitted into an existing fireplace: it can be simply wall-mounted against a suitable chimney breast.

The burners in some back boilers can be noisy. There is also the intermittent sound of the burner cutting in and out on the thermostat—so listen to this type of heater before buying one.

changing fuels

Some boilers can be converted from burning one type of fuel to another. Possible conversions are from solid fuel to oil or sometimes gas; from oil to gas or sometimes solid fuel. (It is not normally possible to convert a gas boiler.)

Conversion costs are likely to be high and it is often more economical, especially if the boiler is more than about five or six years old, to buy a new boiler if you want to change to another fuel. However, you have to add the cost of the actual installing to the cost of the new boiler. With a

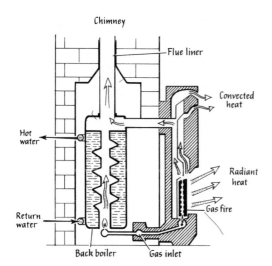

Chimney

Flue liner

Convected heat

Hot water

Radiant heat

Gas fire

Return water

Back boiler

Gas inlet

GAS-FIRED BACK BOILER

conversion or a change of boiler you may also need alterations to the flue. Therefore, to calculate whether it is worth changing fuels, you should compare the likely saving in running costs with the expenses involved in effecting the change.

In dual-fuel boilers with two combustion chambers, you can change from using one fuel to another at will. Some of these boilers burn oil in one combustion chamber and solid fuel in the second. Others are designed to burn solid fuel in one chamber and in the other either gas or oil, but not simultaneously. No dual-fuel boiler which can use gas is approved by the British Gas Corporation.

In some types of boiler, in order to change from one fuel to another, the installer can change the burner.

type of boiler	fuels it can use	advantages	disadvantages
conventional free-standing boiler	gas, oil, solid fuel	long life	takes up floor space
		wide range of heat outputs	needs either a chimney or to be abutting an outside wall for balanced flue
back boiler with a heater in the room	gas, oil, solid fuel	boiler hidden, flames visible	limited range of heat outputs—small to average
		minimum floor space required	location restricted to fireplace or chimney breast
		radiator may not be needed in the room in which the heater is fitted	
wall-mounted boiler	gas, oil	very compact and needs no floor space	limited range of outputs
		rapid heat build-up due to low water content	needs to be abutting outside wall for balanced flue
electric boiler: (*page* 72)	electricity: mainly night rate, or off-peak	no flue or chimney needed	bulky
		no danger of fuel spillage or leakages	limited amount of heat available so may have to be topped up on cold evenings
flow heater (*page* 73)	electricity: mainly day rate, or full price	compact	cost of electricity at daytime tariff
		can be placed anywhere—no flue or chimney needed	

The object of central heating is to achieve a comfortable warmth, so the first priority must be to combat the effect of the cold surfaces that surround us—the walls, floor and ceiling of a room.

Insulation can only do part of the job because there are economic and practical limitations. The rest has to be done by emitting heat into the room in the best possible way. This means counteracting the cold surfaces and heating the cold air in the room, so the heat emitters should be placed close to the coldest parts of the house.

The larger the areas of exposed surfaces, the more difficult is the job of neutralising the cold. In practice, the smaller the area that gives out heat (that is, the more the source of heat is concentrated), the more difficult it is to achieve comfort conditions. So, for instance a long, low radiator stretching along the whole length of an external wall is better than a short, high radiator of identical heat output which covers only a small portion of the wall.

Heat is given out basically in two ways: by radiation and by convection, and all heat emitters do both to some extent. The name radiator is in fact misleading because more heat is generally transferred from radiators by convection than by radiation. With a convector, a small portion of the heat will be radiated.

Convection occurs when air, warmed up by a heated surface, rises by natural buoyancy and colder air flows to replace the warmed air and comes into contact with the heater. This sets up a continuous circulation of air in the room. The air movement can be by natural convection or by forced convection, in which a fan speeds up and directs the movement of air.

When heat is transmitted by radiation, it warms any surface in direct line with the emitter, without appreciably heating up the air between. An open fire heats the people, the furniture and the walls of a room without measurably heating the air in the room. The sun heats the earth across 150 million kilometres of space with its radiant energy unimpaired by the astronomical journey.

In time, the air warmed by a convector and circulating within a room heats objects and surfaces with which it comes into contact. These objects and surfaces then begin to emit heat by radiation. By the same token, objects and surfaces heated by radiation begin to give off some of this heat by convection.

radiators

In this country, radiators are the most widely used type of central heating heat emitter. Inside the metal skin of a radiator are channels through which hot water circulates.

Nearly all radiators used to be made from cast iron (although steel radiators have been made since before the last world war). Cast iron radiators were expensive, bulky and rather ugly. They were quite efficient in use and would last almost indefinitely—some radiators are still in use from the beginning of the century. Cast iron radiators for domestic use have now virtually all been superseded by steel ones, which are much cheaper. However, cast iron radiators, if preferred or if needed to blend in with some existing ones, can still be obtained, though not very easily.

Initially, steel radiators were made to resemble column type cast iron radiators; nowadays they are usually panel radiators, consisting of thin, pressed steel panels, indented with various shapes. These corrugations strengthen the radiator and are designed to increase the heating surface. Panel radiators can be bought in standard heights, from around 300 mm to 750 mm (12–30 inches) and in standard lengths up to about 4 metres (13 feet); you can have larger sizes made to order. The radiators can be bought angled or curved, to suit bay windows for example, at an extra cost and usually with a much longer delivery period.

Heat output from panel and column radiators can be controlled manually or thermostatically; surface temperatures are normally below 80°C.

Extra output is obtained from double panel radiators. These consist of two panels, one in front of the other, with a narrow space between them. (Triple panel radiators are also available.) Such radiators can be bulky and are difficult to clean and repaint. Although there is twice as much panel area, there is not twice as much heat emission, because only the convected, not the radiated heat is increased.

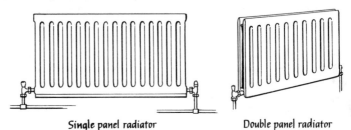

Single panel radiator   Double panel radiator

Other kinds of radiators have been developed which have a higher output per area of wall surface used. These are particularly useful where there is not much wall space. They are single or widely spaced double panels, with metal fins or protrusions on the panels so that there is a larger surface area of hot metal. The extra output is all in the form of convected heat. There are several different patterns of these high output radiators, which are generally more expensive than panel radiators.

A radiator below a window counteracts the effects of the cold surface draughts. Even an undersized radiator placed under the window provides better comfort than an oversized radiator incorrectly located on an inside wall of the room. (Make sure that curtains are short enough not to blanket the radiator.) A shelf above the radiator can help to prevent dirt settling on the wall above, but streaks on either side of the shelf may still appear. Unless specifically designed, a shelf above a radiator will slightly reduce the heat output because it hinders natural convection currents. Boxing in a radiator can seriously reduce its output. Painting a radiator with any metallic paint, such as aluminium or bronze, may also reduce the overall heat output by up to 15 per cent, unless the paint is covered with two coats of clear varnish.

convectors
Natural convectors consist of tubing (through which hot water flows) inside a metal casing. Air enters the open bottom of the casing, passes over the heating tube, rises by natural convection and discharges through a grille or slot in the top of the casing. The heating tube is either plain tubing or has metal fins bonded to it (something like a car radiator), which increases the heating surface and heat output.

A fairly high output can be obtained from natural convectors—higher for the wall space taken up than from ordinary panel radiators. They can, therefore, be fairly small—to fit under a window, for instance. Some models have a damper on the outlet grille which can be opened or closed to vary the output. The output is nearly all in the form of convected heat and surface temperatures are usually quite low.

Fan convectors are the most efficient of any heat emitters from the point of view of high heat output in relation to the size of equipment. They

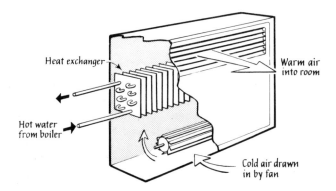

Heat exchanger

Warm air into room

Hot water from boiler

Cold air drawn in by fan

are basically like natural convectors with finned tubing. In addition, there is an electrically driven fan which draws in cooler air, blows it over the heating tubing and out through the discharge grille. The rate at which warmed air is emitted is much greater than with natural convectors. This kind of heater is particularly suitable for rooms where available wall and floor space is restricted or where you need to be able to give a quick boost to the temperature. The fan can be heard, so this type of heater is perhaps better not used for sitting rooms and bedrooms.

Most models incorporate a washable air filter inside the casing which prevents the heater from becoming choked up with dirt.

The fan can be programmed through a time switch to come on and off at pre-set times, and can also be controlled by a thermostat, either in the room or inside the casing. In the latter case, the thermostat senses the temperature of the incoming air stream. Most models also have a thermostat which senses the water temperature in the heating pipe and prevents the fan from operating until the water is sufficiently hot. This prevents cold air being blown out of the heater; it also means that the fan will not be going all night if you have set the temperature low. An override switch is usually fitted so that the fan can be used in summer to provide a movement of air.

For most efficient heat output, the water from the boiler must be very hot. It is better not to mix radiators and convectors (particularly fan convectors) throughout the system because the same water temperature gives different outputs from the different types of emitters. Variations of water temperature would upset the balance of temperatures in the house: any drop in water temperature would result in a greater lowering of the heat output from convectors than from radiators.

Fan convectors are expensive, but very versatile and come in a wide range of different heat outputs. Most models also have a variable fan speed control: the faster the fan speed, the higher the heat output (but the more the noise). This can be very useful for boosting the heating rapidly from a cold start. The speed can then be reduced to decrease the noise. A fan convector should be selected on the basis of its output at normal fan speed and not on the basis of output at boost fan speed.

skirting heating

Skirting heating is fixed along the bottom of the walls of a room in a continuous strip, usually replacing the normal skirting board. For maximum effectiveness, it should be placed, as far as possible, along external walls. It is suitable where, for instance, there are very low window sills so that conventional radiators cannot be used, or where long curtains are required. It is basically an enclosed finned heating pipe, usually less than 225 mm (9 inches) high in all and can be very unobtrusive.

Skirting heating used to be made of cast iron—very durable—either as a radiant heater or as a radiant/convector heater with a high output. Even so, long runs were needed for sufficient heat.

The type installed nowadays is the purely convective skirting heating, which consists of a continuous length of aluminium finned copper tubing (or sometimes steel tubing) in a sheet steel casing. It works like a natural convector with cold air coming in through an opening at the bottom of the casing, being warmed by the heating element and convected out at the top. Dampers are included in the design of the casing to control the heat

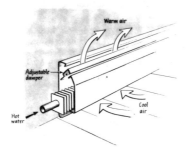

output. The steel casings are not as robust as the cast iron kind, and can be damaged or distorted if knocked. Skirting heating can act as a dirt trap and is difficult to clean. It has a tendency to make a ticking noise when the thermostat turns it on or off, and to creak as the metal expands and contracts.

Skirting heating gives good heat distribution within a room and, in particular, prevents cold draughts along the floor. This means that the room temperature need not be quite as high as for conventional radiators. It is less suitable in rooms with a very high ceiling, above about 3 metres (about 9 feet 6 inches).

room heaters

A room heater can be part of a gas, oil or solid fuel back boiler and is designed to give out a certain amount of heat into the room it is in.

Some models rely mainly on convected heat and others on a combination of convected and radiant heat. If the heater is placed in a large room, larger than its direct output can heat, it may have to be supplemented— by an extra radiator, for example.

The surface temperature of room heaters can be very high, particularly with a solid fuel heater, and where young children or old people are present, a fire guard should be used to prevent accidental burns.

With a solid fuel heater, provided the doors are kept closed, there should be little or no fire risk. Similarly, with a glass-fronted oil-fired room heater, there should be no fire risk from clothing briefly coming into contact with the heater. Most gas-fired room heaters, unless of the purely convective kind, have open flames and although these are shielded with inbuilt wire guards, flammable materials should not be placed too close.

unit heaters

There are gas convectors which can be used individually or linked as part of a home heating system with thermostats and time controls. These convectors have a balanced flue and must be placed against or abutting an outside wall.

The water for an open central heating system comes from an open-topped cistern with a ball valve—the feed and expansion cistern. This is placed at some convenient high point in the house, such as the roof space. The cistern must be a few feet higher than the highest point in the system, for the necessary water pressure. It is more economical in pipework to have it as directly as possible above the boiler.

There is a feed pipe from the cistern to the central heating pipework. Any water lost from the system, due to leakages, draining off or evaporation, is made up from the water in the feed and expansion cistern. With the system cold, the normal water level in the cistern is fairly low and, as the water is heated, it expands up the feed pipe and the water level rises. The cistern should be large enough, therefore, to absorb this expansion so that the water does not discharge through the overflow. Otherwise, fresh water would be drawn into the system each time the water cooled down; this could cause corrosion and, particularly in hard-water areas, scaling up of the pipes, quite apart from wasting water.

sealed system

A sealed system has no open-topped expansion cistern. As water in the boiler is heated, it is allowed to expand into a small closed expansion vessel, generally alongside the boiler. The vessel is cylindrical and on average about 250 mm in diameter and 400 mm high (10 inches × 16 inches). It has a flexible diaphragm inside it across the middle, and air or nitrogen has been pumped in on one side of the diaphragm to a predetermined pressure. As the water expands with heat, the flexible diaphragm is forced out of shape into the pressurised part of the vessel. This increases the pressure throughout the system. With the higher pressure, the boiling point of the water is raised, so that the system can be run at higher than normal temperatures, if required. Such high temperatures are not suitable for radiators, but with convector heaters a sealed system can usefully produce high heat outputs.

With a sealed system, because the water circulates at greater pressure, there is a greater risk of leakage than with an open system, particularly from radiator valves (but special valves can be fitted). You should look at the pressure gauge once a week or so; the slightest leakage shows up immediately as loss of pressure, and must be put right without delay by the installer.

Sealed systems can be used with microbore or small bore pipe systems.

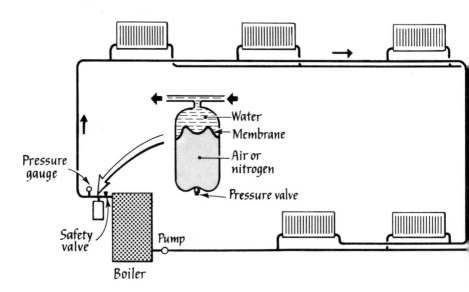

gravity system

This type of piping circuit is very rare nowadays for central heating, but it may be found in older houses, particularly those where the boiler is in the basement or cellar. In a gravity pipe system, water circulates through the pipework and radiators because of the change in its density as it cools. The hotter water close to the boiler is less dense than the cooler water after it has passed through the radiators and given up part of its heat. The denser (cool) water tends to sink in the pipes and the less dense (hot) water is pushed up to replace it.

The normal arrangement is for the boiler to be situated as low in the house as possible and for the hot water to rise directly up to as high a point as possible in the house. The pipes are then routed horizontally across the house to the radiators, and the return pipes descend back to the boiler. There can be several different arrangements, but the general idea is to have the radiators as high above the boiler as possible, and placed so as to have the minimum of horizontal pipe runs. Pipes must be fairly large in order to circulate sufficient water to meet the radiator outputs; the circulating pressure is very low compared to that obtained with a pump.

The biggest advantage with a gravity system is the fact that a pump is not needed and you are therefore less dependent on electricity. (But with most fuels, electricity is necessary for some part of the system to function.) This slight advantage of a gravity system is more than out-weighed by all the disadvantages. The system does not respond quickly to heat requirements because of the long heating and cooling times of such a large quantity of water. The large pipes it needs are expensive to install and the heat loss from them can make the system more expensive to run. You are restricted in where to place the radiators and those near the end of the circuit tend not to get very hot. Fan convectors cannot be used on a gravity system because the narrow water pipes inside the convectors create too much resistance for the low circulating pressure of the water flow.

small bore system

The alternative to a gravity system is one which incorporates a pump to force water through the pipework at fairly high speed (about a metre per second). This means that the pipes can be smaller—usually about 15 to 22 mm in diameter—and their layout correspondingly neater. Small bore pipework is usually of light-gauge copper; light-grade mild steel, which is cheaper than copper, has been used but corrodes very easily and should not be used for central heating pipes.

Because there is not much water inside these small pipes, the system heats up quickly and is much more responsive to controls than a gravity system. The layout of boiler, pipes and radiators is not so critical, as long as any air can be vented from the system. Extra radiators can usually be added to existing pipes without a great deal of difficulty (but the pipework itself can be difficult to extend).

Small bore systems are either basically single pipe or two-pipe. Some heating installations consist of a mixture of single and two-pipe systems.

—single pipe system

In a simple single or one-pipe system, there is a single flow pipe from the boiler which is routed around the house to the radiators or other heaters and back to the boiler. Radiators are fitted above the pipes with connections at opposite ends. Water is pumped through the pipework and enters the radiator through a flow connection and goes back into the pipe through the return connection at the bottom.

In single pipe systems, radiator connections have to be fairly large, particularly for big radiators, because of the low circulating pressure. Successive radiators on the circuit receive water at a lower temperature, and to compensate for this are generally larger.

There are several variations possible in the pipe layout, but the flow and return connections to any one radiator or heater are always from and to the same pipe. An advantage of this system is its relatively low cost: comparatively little pipework is needed and installing it can be quite cheap. But it is difficult to design the system so that it achieves the right balance of radiator temperatures, and single pipe systems are nowadays seldom installed in a house.

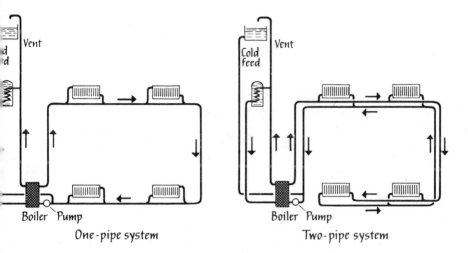

One-pipe system

Two-pipe system

—two-pipe system

In this type of system there are separate flow and return pipes and each radiator, or other heater, is connected to both. Once water has passed through a radiator it does not pass through another but goes directly back to the boiler for reheating. Each radiator therefore has water positively pumped through it and the temperature of the water is roughly the same in all the radiators. The temperature drop within a radiator can be small because of the speed with which the water flows through it, so average temperatures are kept high. This gives a greater heat output for the relatively small amount of heating surface of radiators and is especially useful for convector heaters, where high water temperatures are needed. The flow and return connections are usually both near the bottom of a radiator and quite small.

More pipework is needed with two-pipe systems and when fitted on the surface the two pipes are not quite as neat compared with the single-pipe system. Each radiator needs a special valve to regulate the water circulation through it, otherwise those nearest the pump would take more than their share of hot water and rob radiators farther away.

microbore (or minibore)

There are also two-pipe systems using pipes with very small diameter— 6 mm to 12 mm (as compared with up to 22 mm for small bore). A more powerful pump is needed than for a small bore system. The speed of the water is high enough for any air to be carried through the pipes without creating air locks, so there are no venting problems. Because less water is circulated at any one time, microbore systems respond quickly to temperature adjustments.

Pipework is usually in copper; the long-term life of pipes made of nylon or other plastics has yet to be proved.

The flow and return pipes to and from each radiator lead to a central manifold usually under the floor. From this, there is a flow and return connection from and to the boiler.

Some radiators used in microbore installations have a single connection and valve into which both the flow and return pipes are fitted. This makes the system even neater and quicker to install. Having the correct size pipes is very important with a microbore system—because they are so small, there is literally less room for error. The pipes are often buried in the plaster so it can be very expensive to correct any mistake at a later date.

pumps

An electric pump, which is now almost invariably part of any central heating installation, consumes very little electricity—about as much as a 40 watt electric light bulb—and is quiet in operation. Pumps are small enough to be fitted within a boiler casing, or on the pipework (flow pipe or return pipe) under the floorboards or in a cupboard, but must remain easily accessible for maintenance or replacement. A pump should have a valve on each side so that all the water in the system does not have to be drained if the pump needs to be removed.

Pumps are relatively cheap to buy and it is usual to replace a faulty pump with a new one rather than try to repair it. It is important that the installer fits one capable of pumping sufficient water at the correct pressure around your heating system: not too little, not too much.

A good control system is necessary if you want maximum comfort and minimum running costs. The ideal central heating control system would make you comfortable—neither too hot nor too cold—all the time you are in the house, but without wasting any heat, and without needing any attention from you.

However good a central heating system is, it could be spoilt by bad controls so this is one of the things you should not try to economise on. If you are spending £1100 on installing a central heating system, do not try to save £80 on controls. If that extra £80 would break the bank, decide what controls you would like to have, so that the installer can make provisions for them to be added later. Controls have to be considered at the time of designing your heating system, not after it has been installed.

methods

The simplest method of control—but this is far from ideal—is to turn the radiators down or off by adjusting a valve manually. This restricts or cuts off the amount of hot water going into them, and therefore controls the amount of heat coming out. However, it is almost impossible to be exact in this sort of adjustment, and it is tedious. Alternatively, you could lower the temperature of the circulating water by adjusting the boiler thermostat which controls the temperature of all the water in the system wherever it is to go. However, before a person notices and reacts to a change in temperature and before the radiator has responded to the adjustment, the room will have got even hotter or even colder.

To anticipate your heating requirements automatically, you need something that is going to assess the changes in air temperature in the house even before you have noticed them. Automatic controls react to either a temperature setting or a time setting. Instead of adjusting the boiler every morning and night, for instance, you can fit a time switch which turns the system on and off at pre-set times. A room thermostat can be used to turn the heat off when a room is warm enough and on again as it cools down, without adjusting the boiler. One room thermostat can be used to control the whole house, or you can fit more than one to control zones of the house independently—for example, the bedrooms separately from the living rooms (provided the pipe runs are designed

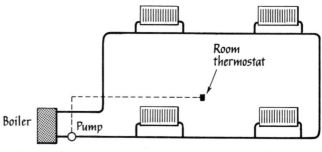

Room thermostat system

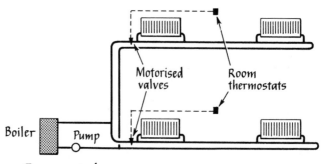

Zone control system

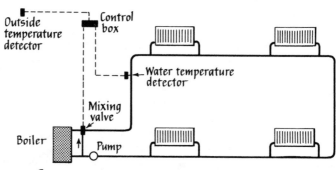

Compensating system

with this in mind). A room thermostat can be used in conjunction with a time switch.

An alternative to room thermostats with time switches is a compensating system. A temperature detector outside the house controls a mixing valve which adjusts the temperature of water in the radiators, according to the outside temperature.

Completely independent room temperature control can be achieved by using a thermostatic radiator valve on the radiators in each room.

The different control systems can be used in conjunction with a time switch or, where domestic hot water is also being controlled, a programmer.

Various combinations of these control systems can be installed to suit your particular needs. For instance, you can have a zone control for a few rooms with the rest of the house governed by thermostatic radiator valves. A fully automatic control system must be capable of detecting and reacting to heat requirements quickly, so that rooms are comfortable at the time you want them to be, and you do not waste heat when, for instance, there is nobody there.

Thermostats

A thermostat is a kind of switch which makes or breaks a connection when a chosen temperature setting is reached. The circuit is broken and the heating stopped at that temperature. When the temperature falls to below the setting, the switch closes, the circuit is again complete and heating is resumed.

Thermostats in central heating systems react to either air temperature in rooms or the temperature of water in the pipes or boiler.

room thermostats

A room thermostat reacts to the temperature of the air immediately around it. It can control the whole heating installation in a house or only the heating of the room in which it is placed. A central heating system is usually designed to give different temperatures in different rooms and if the temperature in one room goes up by a certain amount, it goes up proportionately in the others. If the thermostat controls the heating in the whole house, the room in which it is placed should not be atypical. For instance, if it is the only one where the sun shines in, the thermostat will switch off the heat in the house while the other rooms are still below their required temperature. So where a single thermostat is used, its position is particularly important. If it is put in the entrance hall, it must not be too near the front door where cold air comes in each time the door is opened. A living room is usually not very good because its temperature is likely to fluctuate according to the number of people there and any supplementary forms of heating. In fact, there is no one place that is ideal for the room thermostat.

A single room thermostat, although relatively cheap and simple to operate, is therefore not very satisfactory for controlling the temperatures in the whole house. However, different areas of a house—living and sleeping rooms for instance—can be controlled separately and reasonably effectively with one thermostat for each zone.

Room thermostats should not be placed where air cannot circulate freely around them, such as in an alcove or behind curtains; or too near anything that gives out heat, such as a radiator, TV set or a lamp. An inside wall is better than a cold outside one, as long as it is not warmed by a chimney. The room thermostat should be fixed at about $1\frac{1}{2}$ metres (five feet) above floor level. It has to be wired into the boiler control circuit.

With radiators, the thermostat acts by switching off (or on) the pump or by closing (or opening) a valve in the pipework. Where the pump is turned off, the heat output is reduced gradually: the flow of hot water does not stop straightaway if water can still circulate round the system by gravity. This means that radiators stay warm after the system has been switched off. It is possible to have a null flow or gravity check valve fitted. This stops water circulating through the system by closing when there is no pressure from the pump. Any fan convector in the system must be controlled by its own thermostat (either in the room or inside the casing), which acts by stopping or starting the fan rotating. The reaction is instantaneous.

If the boiler is not used for heating the domestic hot water supply, a room thermostat can act directly on the boiler by switching the gas or oil burner off (or on) and with a solid fuel boiler by operating the damper or draught fan.

Room thermostats can be set to a low setting to bring the heating on when the temperature gets very low—to prevent frost damage in an unoccupied house. Alternatively, a second thermostat (a frost thermostat) can be fitted in addition to the normal room thermostat, preferably in the coldest part of the house, and wired so that it will override all the other controls and switch the heating on for a time if the temperature falls to a very low level.

A night set-back that sets a lower temperature at night than in the daytime is incorporated in some room thermostats. This lower setting can be selected either manually by setting a timer each night or, with some models, by using a built-in time switch which changes the temperature settings automatically down at night and up in the morning. The same result can be achieved by having two thermostats set to different temperatures and a time switch.

A house in which the daytime temperature has been consistently high often does not cool down to the night set-back level because some heat will have been retained in the fabric of the house.

boiler thermostats
Immersion thermostats are used for sensing water temperatures and are fitted into the water jacket of a boiler, or directly into a pipe.

When used as boiler thermostats, which is their normal function, they are set to switch off the oil or gas burner at a selected temperature. Natural draught pot boilers cannot re-light themselves automatically, so are not switched off completely: the thermostat reduces the oil flow to give the lowest amount of heat.

The maximum and minimum settings for domestic boilers should not be above 80°C (176°F) and below 60°C (140°F). Above 80°C the radiator and pipes would normally be too hot if you touched them accidentally. Below 60°C, there is a likelihood of corrosion in the boiler. The thermostat adjusting knob on the boiler is graduated either in degrees or in numbers, for instance 1 to 9, where the range of temperatures represented by the figures could be from about 50°C to 90°C If you have a thermometer in the boiler you can check that the thermostat is working correctly.

A boiler thermostat alone is not enough if the boiler heats water for the domestic hot water supply as well as the heating. Without independent control of each, the domestic hot water may at times be too hot.

Solid fuel boilers cannot be controlled in the same way as gas and oil boilers by automatically cutting off the fuel. Instead, the immersion thermostat operates a damper in the air inlet, or a fan in the boiler.

—limit thermostat

A limit thermostat may be fitted to a boiler as an added precaution against the water overheating or boiling. It is simply another boiler thermostat set at a higher temperature—say 90°C compared with 80°C for the normal boiler thermostat. If for any reason the boiler thermostat fails, the limit thermostat should come into operation and turn off the heat source. Some limit thermostats have a manual reset button which must be pressed to bring the heating back on—this draws one's attention to the fact that something is wrong. Other limit thermostats are thermally reset and will automatically switch on the heat source once the temperature has dropped below the set point.

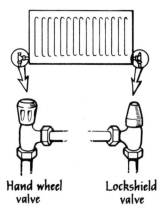

Hand wheel valve   Lockshield valve

Valves

A valve is a device to stop or control the flow of liquid or gas through a pipe. The simplest, in a central heating system, is the hand wheel valve on a radiator.

hand wheel valves

You operate a hand wheel valve simply by turning the knob at the top. Nowadays this is usually made of a plastic material; the body of the valve is metal. A hand wheel valve is normally fitted on the flow connection to each radiator (or other heat emitter).

Hand wheel valves can also be used in other parts of the heating pipework, for example on each side of the pump so that it can be isolated for repair or renewal.

lockshield valves

Lockshield valves are similar in appearance to hand wheel valves, but to get at the spindle for adjustment, the plastic cap has to be removed. Lockshield valves are fitted on the return connection from a radiator (or other heat emitter). The initial adjustment has to be made by the installer with a special tool or screwdriver. The valve regulates the flow of water through a circuit or radiator so that the water flow throughout the complete heating system is balanced.

When fully closed, such a valve can be used as an isolating valve together with a hand-wheel valve on the other side of the radiator when, for example, a radiator is leaking or is temporarily removed for decorating.

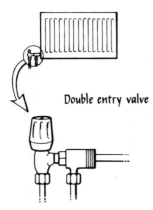

Double entry valve

double entry valves

A single valve which incorporates the actions of the hand wheel and the lockshield valves can be fitted to radiators, either on one side or, with radiators specially designed for it, in the centre at the bottom of the radiator. The flow connection and return connection are both through the double entry valve. These valves are often used in microbore systems.

thermostatic valves

A thermostatic valve can be fitted on a radiator or heater in place of a hand wheel valve, directly on the radiator or in the flow connection. These valves are designed to react to the air temperature immediately around; when this rises above a certain level, an expansion mechanism in the valve cuts down or closes off the supply of hot water to the radiator.

This kind of valve is completely non-electric and can be set to close off at any desired room temperature. (Settings are in figures or symbols and you may need to find the right one for the temperature you want by trial and error.) They are not particularly expensive when you consider that the price of a hand wheel valve is saved, plus the probable cost of an alternative control system. With some thermostatic valves you can remove the thermostatic head and replace it with a cap so that the radiator can be removed—for re-decoration, for instance—without the whole system having to be drained. But with others, it is a wise precaution to have a small plug-type valve as well as the thermostatic valve, for when a complete shut-off is needed.

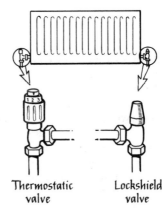

Thermostatic valve

Lockshield valve

Thermostatic radiator valves can be fitted on any size pipe. However, they are not widely available for the very narrow pipework of microbore systems.

By using thermostatic radiator valves, the temperature of each radiator—and therefore of each room—is controlled individually. This means that rooms which are used only occasionally, such as the spare room, can be kept to a low background heat. If one room gets unusually hot—for instance if there is a party—the thermostatic valve in that room adjusts the temperature without cooling down the rest of the house.

Another advantage of thermostatic valves is that their action is continuous. They gradually increase or decrease the water flow through the radiator. The heat is modulated rather than stopped and started and the effect is smoother and more comfortable.

You can have an existing hand wheel valve replaced by a thermostatic valve even after an installation has been completed.

two-way motorised valves

A room thermostat can control the flow of hot water through the pipework by operating a two-way (open/shut) motorised valve.

For zone control, different parts of the house are controlled by separate room thermostats acting on two-way motorised valves in the pipework which serves different parts of the house. A time switch can be incorporated.

three-way mixing valves

A mixing valve lets cooled water returning from the radiators mix in with the hot water flowing from the boiler, so that the water reaching the radiators is cooled to the required temperature without the temperature of the water going to the domestic hot water cylinder being affected. This would be useful, for instance, in mild weather when you need minimal heating but want the water to be hot for baths. With a manual mixing valve, you adjust a knob or lever; a motorised mixing valve operates according to the demands of a thermostat or other switch, or is part of a compensating or modulating system.

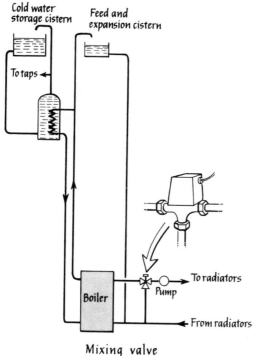

Mixing valve

A three-way mixing valve is not normally used in conjunction with fan convector heaters because their heat output falls off too rapidly with a fall in water temperature.

Compensating systems

A weather compensator is made up of a control box, an outside temperature detector which activates a motorised mixing valve and a water temperature detector. This detector monitors the temperature of the water flowing to the radiators, and if it is too hot or too cool, the mixing valve is further adjusted by the control box.

The outside detector should be mounted on an external north-facing wall, in free air because it is sensitive both to air temperature and to wind effects. The control box can also incorporate a time switch to give a reduced night time setting and sometimes an early morning boost.

Such a system is fully automatic and does away with the problem of which room to put a thermostat in. However, it does not respond to temperature changes in a room brought about by, say, a lot of people being in a room (as individual thermostatic radiator valves do).

Instead of the outside temperature detector, some systems have a room temperature detector, to operate the mixing valve. As with a room thermostat, the problem of where to place the temperature detectors remains. The advantage of this system over a room thermostat is that the heat change is modulated rather than producing an abrupt on/off effect.

A very sophisticated compensating system is one which incorporates both an outside and a room temperature detector.

Modulating control systems are not suitable with fan convectors.

Programmers

A programmer is basically a time switch designed to provide different heating programmes. By switching the pump on or off, or by operating one or more motorised valves, heat can be diverted or given preference either to the heating circuit or to the hot water supply. These preferences can be preselected or programmed, and timed to operate at different periods of the day or night. A programmer can either be a small unit fixed to the wall, or it can be an integral part of some makes of boiler.

When installing central heating, you may well already have a hot water system. This can either be left completely separate from the new heating system or incorporated into it. If there is no existing hot water system, or you do not want to use it any longer, a combined central heating and hot water system can be installed from scratch.

You can also have an electric immersion heater fitted into the domestic hot water cylinder for summer or emergency use.

If you are going to use the existing hot water system, the cylinder will have to be converted into an indirect one, that is one where hot water from the boiler heats tap water inside the cylinder without the two waters mixing. Alternatively, you must buy and fit a new indirect cylinder.

circulation

The circulation of water from the boiler to the domestic hot water cylinder can be by gravity or pumped, irrespective of the circulation system in the central heating.

Where circulation is by gravity, the cylinder should be located as directly and as high above the boiler as possible to give a good circulation. The distance from the cylinder to the taps should be as short as possible to minimise the amount of cold water dead leg, otherwise taps have to be left running for some time before the hot water comes through and a lot of water is wasted.

Where the circulation is by pump, the location of the cylinder is less critical and the pipes from the boiler to the cylinder can be smaller (and are therefore cheaper). The pump can be either the one which pumps the hot water through the central heating part of the system or a separate one.

With a solid fuel boiler (or a natural draught pot boiler) the circulation to the cylinder must be by gravity—so that the cylinder can absorb excess heat from the boiler. This is because the fire takes some time to burn out after the pump has stopped circulating water to the radiators. The heat therefore has to be able to rise to the cylinder and there must be no valves between the boiler and the cylinder because they could cut off the flow. It is an added safety precaution to have one radiator in the system (perhaps in the bathroom) to which the water circulates by gravity.

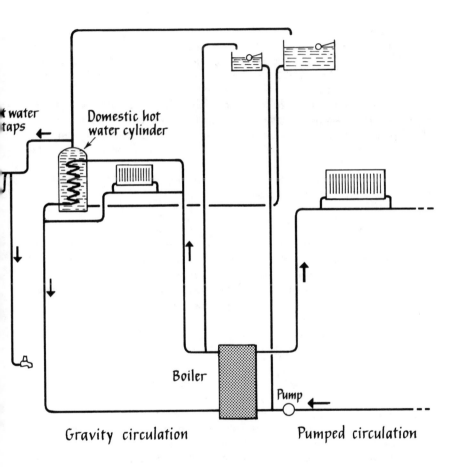

◀ water taps

Domestic hot water cylinder

Boiler

Pump

Gravity circulation

Pumped circulation

If you have a heated towel rail, it is sensible to connect it to the pipework which carries hot water from the boiler to the cylinder. This means that the towel rail can be heated when the radiators are not—in summer, for instance.

cylinders

Domestic hot water cylinders are usually made of copper or sometimes galvanised steel. Cylinder capacities vary; the present-day standard size is around 125 litres (27$\frac{1}{2}$ gallons), but a larger one can be installed if there is enough room and heating capacity.

A cylinder should be insulated, otherwise the heat loss will be considerable. An 80mm insulating jacket is recommended. A cylinder in an airing cupboard gives off enough heat to keep clothes and linen aired even when it has an insulating jacket. Pipes connecting the boiler to the cylinder should also be insulated.

There are three different types of cylinder. The first (direct cylinder) is hardly ever installed nowadays.

—direct cylinder

With a direct system, there is only one cold feed cistern. The water which is heated in the boiler is the water which flows through the radiators and also runs out of the taps, with no intermediate heat exchanger.

This arrangement is largely obsolete and very unsatisfactory because the constant introduction of fresh water into the boiler and pipework inevitably leads to scaling and corrosion. However if you have an independent domestic hot water boiler, separate from the central heating system, there is no reason why a direct cylinder should not be used.

—indirect cylinder

An indirect cylinder looks like a direct one from the outside. However, inside there is a heat exchanger (either a copper coil or an annular ring) through which very hot water flows from the boiler. This heats up the water around it, in the cylinder, which is destined for the taps. The two waters do not mix. Scaling is not a problem because fresh water is not constantly being introduced into the heating water.

A direct cylinder can sometimes be converted into an indirect one by inserting a heat exchanger. Provided the cylinder is in good condition, this can be cheaper than having a new cylinder when central heating is being installed. However, this is not recommended with a galvanised steel

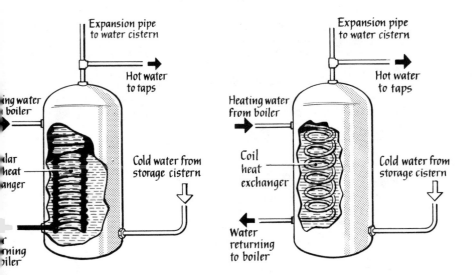

Expansion pipe
to water cistern

Hot water
to taps

ing water
boiler

lar
heat
anger

Cold water from
storage cistern

ing
iler

Expansion pipe
to water cistern

Hot water
to taps

Heating water
from boiler

Coil
heat
exchanger

Cold water from
storage cistern

Water
returning
to boiler

INDIRECT CYLINDERS

cylinder because the mixture of metals involved would lead to corrosion.

Two separate cisterns are necessary: one the feed and expansion cistern, the other the cold water storage cistern for the house. They are usually fixed alongside each other in the roof space. Both also act as expansion cisterns.

—automatic self-feed cylinder
There is a variation of the indirect cylinder system, which uses an automatic self-feed cylinder for which only a storage cistern and no feed and expansion cistern is needed. (This means less pipework and therefore less money.)

Any expansion from the heating part of the system flows into an expansion chamber which is part of the heat exchanger inside the cylinder. The expansion chamber contains a cushion of air which separates the heating water and the tap water. It can take only a certain amount of expansion from the heating system, and if overloaded the air cushion would become misplaced and allow the heating and pipewaters to mix. It is therefore important that the cylinder is suitable for the heating installation. It cannot be used with a sealed system, the essence of which is a slightly raised pressure: this cannot be attained in a system with an automatic self-feed cylinder.

There must be no valves of any kind between the boiler and the self-feed cylinder.

cylinder controls

To control the temperature of the water in a domestic hot water cylinder, a contact thermostat can be used. It is fitted on the outside of, but in contact with, the cylinder, and opens or closes a motorised valve which regulates the flow of water from the boiler to the cylinder. Where the circulation is not by gravity, the thermostat controls the pump that circulates the water to the hot water cylinder. With a cylinder thermostat, the boiler can be run at a temperature that is high enough for the radiators while the domestic hot water can be kept at a temperature low enough not to scald you when it comes out of the tap. Boilers should be run at above 60°C (140°F) to avoid corrosion. The cylinder thermostat setting should be kept below 60°C: the cylinder is more likely to scale up at a higher temperature, particularly in hard water areas.

During the summer, when the boiler is used only for heating the domestic hot water, the cylinder thermostat can switch the burner off when heat is not required (instead of stopping the water flow); a separate switch is necessary for this.

A cylinder thermostat can also be used with a diverting valve. This is a three-way valve, fitted where the water that is going to the heating system diverges from the water going to the domestic hot water cylinder, so that the valve can automatically shut off one or other of the two supplies. (With some valves, a mid-position allows some hot water to flow to both.) The diverting valve is governed by the cylinder thermostat and a room thermostat, in conjunction with a programmer.

Another way of controlling the temperature of the domestic hot water is to have a thermostatic control valve fitted in the pipe from the cylinder, so that it senses the temperature of water returning to the boiler. When this rises above a certain temperature, the valve automatically closes, so that the circulation of heating water to the cylinder stops. The valve incorporates a hand adjustment so that the switch-off temperature can be varied.

No cylinder control valve of any kind should be used with a solid fuel boiler or a vaporising oil burner where the heat is not completely shut off when not required. If all the radiators are turned off and then the cylinder were also to be turned off, there would be no outlet for the residual heat. This would ultimately raise the temperature of the water in the boiler to boiling point—which could cause considerable damage to the boiler.

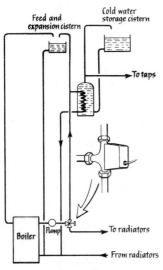

Diverting valve

It is usually not practicable to install a warm air system into a house that is already built. But if your house is still being built, a warm air system is an alternative to a wet system. The main difference between the two is that warm air instead of hot water is used to heat the rooms.

The system
The air is heated as it passes over the heated core of a warm air unit and is distributed through ducts to grilles in the rooms that are being heated. Unlike wet systems, which have a flow pipe bringing hot water to the heat emitters and a return pipe through which cooler water returns to the boiler, in a warm air system there are no return ducts from each room back to the warm air unit. The air is drawn into the unit by the same fan which blows the hot air through the ducts. Before being re-circulated, the air is filtered in the unit, but smells—be it cigarette smoke from a party in the lounge or onion from a fry-up in the kitchen—tend to get blown round the house. It is sometimes possible to lead a duct from the open air to the heating unit through which fresh air can be introduced into the system.

units
The majority of warm air units are direct-heating: the heat is transferred directly to the airstream without first heating up water as an intermediary, so there is no heating-up delay.

A warm air unit is a self-contained heater complete with burner and controls, an electrically-driven fan which blows air through or over the heat exchanger and into ducts, and usually an air filter.

Warm air units are usually tall and some extend from the floor to the ceiling.

Warm air units are designed to be run on gas, oil or electricity. A gas or oil warm air unit needs a similar flue to a boiler.

A warm air unit run on electricity (Economy 7 or white meter) does not need a flue or chimney. It consists of a central core of either metal or refractory brickwork, heated by electricity during the night. On very cold days, temperatures may tail off towards the evening but can be boosted by heating up the unit by day rate or full price electricity.

There are also warm air units which give out heat indirectly to the air stream: the heat from the combustion chamber is first transferred to water surrounding it, and then from the water jacket to air which is ducted round the house. It takes a little longer to heat up, and there is therefore more delay between starting up the unit and heat coming out of the grilles than with a direct warm air unit. However, the water jacket round the combustion chamber is a kind of protection: with a direct unit if the combustion chamber were to burn through, there would be the risk of products of combustion getting into the air stream.

The hot water in an indirect warm air unit can also be used to heat up the water in a domestic hot water cylinder. With a direct warm air unit, a separate water heater can sometimes be incorporated into the casing of the unit.

A warm air unit can be fitted in a convenient place not suitable for a boiler—generally it is better placed in the centre of the house, to make the ducting less extensive.

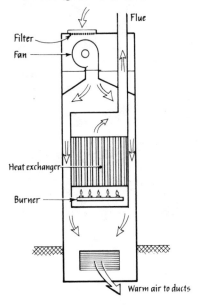

Filter
Fan
Flue
Heat exchanger
Burner
Warm air to ducts

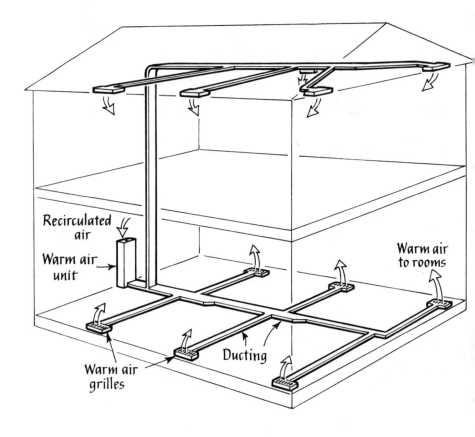

Recirculated air

Warm air unit

Warm air to rooms

Ducting

Warm air grilles

ducting

The ducting which carries warm air from a heating unit to grilles in the rooms is generally concealed in the roof space and under the floor. Where it runs above ground in rooms, it is encased and can be painted or papered over.

Ducting is either circular or rectangular and can be made from fibre-board, glass fibre and some plastics, but is mostly made from galvanised sheet steel. Ducts can be bought ready-made complete with fittings, such as bends or tees (which join sections at right angles to form a T-shape). Alternatively, the ducting can be specially made to suite a particular kind of installation.

Because of the relatively large surface area of ducting, all ducting should be well insulated—especially if made of metal. Some metal ducting can be bought coated with a foam type of insulation bonded on to it. Good insulation (a minimum of 50 mm, 2 inches) is particularly important where the ducts are under the ground floor or in the roof space.

Any one installation will have a variety of duct sizes, usually getting smaller after each division away from the unit. Your designer must calculate the dimensions of the various ducts carefully to make sure that the right amount of air is carried with the least noise, least heat loss and least cost. Also, because sound travels very easily through ducting—which acts like a speaking tube—care must be taken with the layout of the various ducts and grilles so that the noise from a television set, for example, is not transmitted directly to a child's bedroom. If the relationship between the size of the unit's fan, the ducts and the grilles is incorrect, the system will be noisy or not warm enough for comfort.

The cool air has to be circulated back to the heating unit to be warmed. It gets out of each room by return grilles. These should be at low level, regardless of where the inlet grilles are. Inlet and return air grilles should not be close together in a room, to prevent short circuiting of warm air directly from the inlet to the return grille.

Some units have a duct leading from the outside to the unit through which fresh air goes straight to the unit. In the winter, the damper in the fresh air duct is normally kept almost shut. In the summer, it may be kept fully open and if the fan alone is switched on (without the heater) fresh air is circulated around the house. Without an outside duct, the fan can also be used in the summer to provide air circulation, without bringing in fresh air. It may be possible to add a fresh air duct after the unit has been installed but it is better incorporated in the initial design.

grilles

The warm air is discharged into the room through grilles in the ceiling or the floor or walls. In a large room, more than one inlet grille will be needed. Although grilles are less obtrusive than radiators, and generally smaller, they can interfere with carpets and nothing must be put in front of or over a grille. Grilles made of steel can be painted any colour; anodised aluminium grilles are mostly left unpainted.

Since warm air rises, inlet grilles at low level give better heat distribution. Grilles at a low level in the wall below the window help to counteract downdraughts. A practical arrangement is to have the grilles on the ground floor near the bottom of the wall (or in the floor) so that the ducting can be under the floor; and near the top of the wall (or in the ceiling) on the upper storey of the house, to allow the ducting to be lead through the loft.

controls

A room thermostat can be used which switches the burner or the circulating fan in the heating unit on or off. A time switch can be used to make the room thermostat operative at certain hours. When the fan is switched off (or on) the reaction—that is the change in temperature—is rapid. With an indirect warm air unit, the thermostat acts on the fan only, so that domestic hot water can still be produced.

Dampers in warm air ducts are either hand-operated or motorised. Motorised dampers can be opened or closed according to the demands of a room thermostat to which they are wired. Being fairly expensive, motorised dampers are used to control a whole floor or a zone, usually in conjunction with a time switch, rather than being fitted on an individual basis for each room. There is no practical way of controlling individual rooms automatically except going to the great expense of fitting an automatic damper to the grille in each room. But in rooms where you want to be able to control the temperature, you can have grilles with manual control dampers, and adjustable blades with which the direction of air flow can be varied. Dampers can also be used to shut off grilles in individual rooms so that all the heat is concentrated into the other rooms, for a rapid build-up of heat there. However, shutting off some grilles can increase the air velocity through others and so creates more noise.

stub ducts

It is easiest to install warm air central heating while a house is being built. With an existing house, the installation of lengthy runs of ducting could necessitate structural upheavals and much expense. However, it is some-times possible to position a warm air unit centrally in a house, with very short ducts—about a foot (300 mm) long—leading to grilles in the rooms. Such a stub duct unit would give full heating downstairs and only background heating upstairs. As with all warm air heating, there is no radiant heat and because with stub ducts it is impossible to place a grille below the window, there is no heat source to counteract cold radiation from the window.

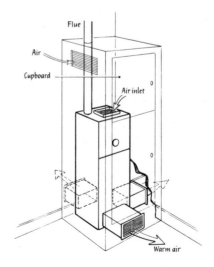

humidity control

If your central heating makes the air seem uncomfortably dry, you may be able to have a humidifier fitted into the ducting or the unit, so that the fan blows air over damp pads. There are other humidifiers, such as electrically-operated ones which come on automatically if the humidity falls below a certain level, and can also be operated manually. The humidifier works by spraying very fine particles of water into the air. It can be put anywhere in the room.

Electric heating using off-peak electricity makes it possible to take in electricity at the time of the day when it is cheaper and storing the heat it produces for use later. It is possible to boost night rate heating by taking in day rate electricity when it is needed to increase the output.

In so-called direct acting systems, heat is produced for immediate use rather than stored, and the tariff for the electricity used depends on the time of day.

Electric central heating systems do not include provision for heating domestic hot water. An immersion heater in the domestic hot water cylinder can be run on off-peak electricity—but the cylinder has to be well lagged.

off-peak electric storage heaters

Thermal storage heaters use electricity at off-peak times: the heaters take in heat during the night only and release it gradually. The rating of a storage heater refers to the rate at which electric current is taken in, not the heat output—which is lower.

These heaters should never be covered with anything—however tempting it may be to put laundry to dry or linen to air right over the heater. This would cause a rise in the temperature of the casing so that materials could be scorched. To prevent overheating, storage heaters have a thermal link inside the casing which melts at a certain temperature and so cuts off the electricity supply. Even when the supply has been cut off, the casing temperature continues to rise because of the heat still inside the refractory core of the heater. Before the heater can be used again, a correct replacement link must be fitted by an electrician.

With the simplest type of storage heater, only the amount of heat taken in during the night can be controlled, not the amount given out. So on a particularly cold day there may not be enough warmth left by the evening, while on an unexpectedly mild day you may have to open the windows.

There are also off-peak storage heaters in which the heat output can be regulated by a damper which allows the flow of convected air when opened; but this is limited to the end of the discharge period.

More sophisticated models have a built-in fan. Heat output from them is mainly convective. When the fan is switched off, very little heat is released. The heater's fan can be operated thermostatically so that it will switch on when the room temperature falls below a certain level. It is perhaps worth the extra expense because heat is not used up when it is not needed and saved for when it is, such as the evening.

If there are enough storage heaters in your house, they can provide full central heating, but they are often used as a source of background heat to be boosted to full comfort by additional forms of heat, such as standard or day rate electric heaters, or gas fires.

The heaters are bulky but, in theory, can be moved—to another house when you leave, for instance. But this would involve dismantling and then rebuilding them. If you want to add more heaters in a house, this can be done simply, without any structural alterations, as long as the wiring is adequate, and the floor strong enough.

underfloor heating

Electric underfloor heating is suitable for use with off-peak supply because the heat from the elements embedded in concrete floor slabs is retained and emitted gradually. However, the rate of emission cannot be controlled. Underfloor heating cannot be installed in an existing house unless new flooring is being put in (nor in a house with ventilated wooden floors). The upper floor in a house will probably need some other form of heating.

electric boilers

An electric system using off-peak or white meter night rate, consists of a large well-insulated vessel—from about 450 litres (about 100 gallons) capacity—heated by immersion heaters and from which water is pumped around a normal radiator system. Hot water is pumped from the vessel only when heat is needed. Provided the water in the storage vessel is kept hot, the radiators respond very rapidly when heat is needed. This system can be adapted to heat the domestic hot water, but a separate domestic hot water system is generally more efficient.

The units are rather bulky, but there is no need for fuel storage and also no need for a flue, so they can be put anywhere, for instance in the cellar or under the stairs, or the garage.

The floor on which the vessel stands must be capable of supporting the very heavy weight—some of the larger models can be over 1,600 Kg (about $1\frac{1}{2}$ tons). These water thermal storage heaters are not very common and you need to ask your electricity board where you can order one.

flow heaters

An electric boiler, similar in size to a wall-mounted boiler of comparable heating capacity, can be used to heat water with immersion heaters that use day rate electricity (or night rate, according to when it is running). Water is being constantly re-heated as needed, and is circulated through radiators or convectors, as in any other wet system.

ceiling heating

Like underfloor heating, ceiling heating is completely out of sight and takes up no room, but is direct acting and does not store heat, so has to use full price or day rate electricity and takes in off-peak electricity only when it uses it at night.

The most commonly used form consists of plastic sheets that contain heating elements, fixed in continuous lengths on to the ceiling joists or battens. There is a layer of insulation above the heating sheet and plasterboard or hardboard below, which can be decorated in the normal way. As the elements heat up, warmth is radiated downwards into the room and heats all parts evenly. A thermostat in the room switches the ceiling heating on and off according to the temperature, and can operate through a time switch.

panel heaters

Metal panels containing electric elements are direct acting and mainly wall-hung—convectors or radiant convectors. They can be controlled by individual room thermostats and a central time control. They are similar to ordinary individual electric heaters which you plug into a wall socket. However, they are wired into a separate circuit and the time control switches the heaters on (and off) in groups of rooms at different times. They need to use day rate or full price electricity. Some metal panels have a thermal cut-out as a safety device: avoid those that do not.

Panel heaters with plastic rather than metal panels can be a fire risk and should be avoided where there is any risk of their being covered or of having furniture placed too near.

If you buy a BEAB—approved make of heater it has been tested to the relevant British Standard and carries the BEAB mark. The British Electro-technical Approvals Board is concerned with the safety aspect of electrical appliances used in the home. Its approval scheme covers both british-made and imported equipment.

costs

The installation cost of electric central heating which does not have pipework or flues should be comparatively inexpensive. Direct acting systems are very controllable—there is hardly any maintenance, no fuel needs to be delivered or stored. But all these advantages are likely to be outweighed by the cost of running any system that uses full price electricity. They are promoted as being economical if you have good insulation and a good control system, but any central heating is more economical with good insulation and controls.

THE INSTALLING

If money is no object, a consulting heating engineer will design a central heating system for you (from deciding which fuel to use, to where to put the last air vent), find the installer and supervise his work. You pay him either a flat fee or a percentage of the total cost. A consultant may work on his own, or in a partnership or group.

Alternatively, you can use someone who will design the system and also carry out the installation. You can find him through one of the professional or trade organisations who will give you the names of their members.

Or you may seek professional advice on the design of your central heating and then find yourself an installer and supervise him.

The commercial, semi-commercial, professional or trade organisations can give general and specific advice about fuels and central heating, and will possibly help you find an installer. If you have already decided which fuel to use, you can approach one of the three nationalised fuel industries for information, or the major oil firms.

Chartered Institution of Building Services

If you want to engage a consulting engineer to design and supervise your central heating installation, perhaps because the building poses special problems, you can write to the secretary of the CIBS, 222 Balham High Road, London SW12 for names and addresses of one or two of the institution's members who are in consultant practice.

National Association of Plumbing, Heating and Mechanical Services Contractors

The address of this trades association is 6 Gate Street, London WC2A 3HX, tel 01-405 2678. There are regional offices in Leeds, Rochdale and London SE18 which can give you the names of association members who carry out central heating work in your locality and advise on the type of installation suitable to your circumstances. The association has a code of fair trading for work carried out by its members; it includes an arbitration procedure.

The Heating and Ventilating Contractors' Association

The HVCA is a trade organisation (at Esca House, 34 Palace Court, London W2 4JG). Their Home Heating Group operates a customer enquiry service. You can telephone their home heating enquiry line on 01-229 5543, or write, and they will give you the names of two of their members who can install central heating in your area. The HVCA member will give you advice on the choice of fuel and system. If you have the installation done by him, you get the HVCA double guarantee—a one-year guarantee of workmanship, materials and performance. The 'double' part means that if the installer does not honour the guarantee, the HVCA will do so. The HVCA will deal with complaints about their members.

Publications from 10 King Street, Penrith, Cumbria, CA11 7AJ.

Building Centre

The Building Centre (26 Store Street, London WC1E 7BT, tel 01-637 9001, includes an exhibition of central heating equipment from a wide range of manufacturers whose literature relating to the displays can be taken away by people who call in person; postal enquirers should enclose 20p towards postage.

Within the Building Centre there are also permanent exhibitions maintained by British Gas Corporation and — with staff in attendance — the Electricity Council and Solid Fuel Advisory Service.

There are also Building Centres in Birmingham, Bristol, Cambridge, Durham, Glasgow, Liverpool, Manchester, Nottingham, Southampton.

Institute of Domestic Heating and Environmental Engineers

The Institute (93 High Road, Benfleet, Essex, tel 03745-54226) publishes a list of members engaged in contracting and has a consultancy group of members who give advice, on a fee paying basis. The IDHE publishes a standard specification for central heating installations, available to contractors and customers.

The Institute of Plumbing

Many of the Institute of Plumbing's members are self-employed plumbers or directors of plumbing companies who install central heating. The Institute (Scottish Mutual House, North Street, Hornchurch, Essex RM11 1RU, tel Hornchurch 51236) will provide names and addresses from its register of plumbers, but does not guarantee any work carried out by them. The Institute of Plumbing will investigate complaints against members.

Builders Merchants Federation

The BMF is a trade organisation, many of whose members sell central heating equipment. You can get a list of authorised dealers from the Federation (15 Soho Square, London W1V 5FB, tel 01-439 1753). Although the Federation does not officially recommend installers, its members often have useful local knowledge and experience. They display and give advice on central heating equipment in most of the BMF's home improvement centres.

Electricity

To get general information about electric central heating, you can go to your local electricity board shop. They will be able to give you a very rough idea of how much an electric central heating installation would cost you. You should make an appointment for a technical adviser from the board to come to your home and give a more detailed quotation later. The board is likely to have its own employees who can carry out the work, but may subcontract all or part of the work. The contract and guarantee, however, remain with the electricity board. Electricity boards offer servicing schemes.

Every area electricity board has an electricity consultative council, whose job includes investigating complaints from consumers which have not been settled satisfactorily by the local electricity board. You can get the address from your local electricity board shop.

Instead of paying quarterly accounts, electricity boards accept payments in equal monthly instalments. The amount you pay is based on estimated consumption and is adjusted once a year.

National Inspection Council for
Electrical Installation Contracting

APPROVED CONTRACTOR

The National Inspection Council for Electrical Installation Contracting (NICEIC), 237 Kennington Lane, London SE11 5QJ, tel 01-582 7746 publishes annually a roll of approved electrical contractors who have undertaken to comply with the regulations for the electrical equipment of buildings (the IEE wiring regulations) and whose work is subject to regular inspection by the NICEIC.

The NICEIC will investigate complaints relating to the safety of an electrical installation by one of its approved contractors, but does not adjudicate on the effectiveness of electrical heating systems not covered by the IEE wiring regulations.

Copies of the roll are at electricity board shops, consumer advice centres and CABs, the names and addresses of approved contractors may also be obtained from the NICEIC. At present, there are about 8,000 names and addresses on the roll, including electricity boards and most members of the Electrical Contractors' Association.

The Electrical Contractors' Association (34 Palace Court, London W2 4HY, tel 01-229 1266) will supply names and addresses of their members willing to advise on and carry out electric central heating installations. Their work complies with the regulations of the Institution of Electrical Engineers. The contract is with the installer, but his workmanship carries the ECA guarantee. The ECA will deal with any complaint against a member about installation work that has not been resolved by him. There is also a completion guarantee scheme if the contractor goes bankrupt.

In Scotland, a similar service is offered by the Electrical Contractors' Association of Scotland, 23 Heriot Row, Edinburgh EH3 6EW.

Gas

The British Gas Corporation (59 Bryanston Street, London W1A 2AZ) consists of the 12 gas regions, previously called gas boards. For general advice and information on gas central heating, you should go to a local gas region showroom. The gas region will either carry out the installation or contract the work out to an approved installer.

British Gas guarantee their advice as well as their equipment and installation. They have various types of service contracts. If you have a complaint that cannot be dealt with by the gas region, you can take the matter up with your regional gas consumers' council whose address can be obtained from the showroom.

If you want to make your own arrangements direct with an installer, the gas showroom will let you see a copy of the CORGI register (which includes details of what each firm specialises in). CORGI stands for the Confederation for the Registration of Gas Installers, whose main concern is the safety of gas installation and servicing work done by private contractors. CORGI operates from St. Martin's House, 140 Tottenham Court Road, London W1P 9LN, tel 01-387 9185, and has 12 regional offices, details of which can be obtained from London.

You can pay the gas bill for central heating in monthly instalments based on an estimate of how much you will be consuming. The amount is adjusted at the end of each year.

Solid fuel
The Solid Fuel Advisory Service (Hobart House, Grosvenor Place, London SW1X 7AE, tel 01-235 2020) has regional and local offices; gives free advice on fuels, appliances and systems and can arrange for a representative to visit your home and discuss your heating requirements.

You can get a list of SFAS registered heating contractors from Hobart House or regional or local offices. Your contract is with the contractor whose two year guarantee covering workmanship, performance and materials is backed by the SFAS. The SFAS also lists approved appliance installers who are outside the SFAS two-year guarantee scheme.

SFAS approved appliance distributors advise on central heating equipment which they sell, and can provide lists of registered heating contractors and approved appliance installers.

You can also get a list of the SFAS living fire centres which are showrooms and information centres.

As well as at its own address, the SFAS may be contacted through coal merchants and through the National Coal Board.

Oil
Services offered by the major oil companies to people having central heating installed range from just delivering oil to advising on, installing and maintaining the whole system: it is worth contacting the local oil company office or authorised distributor for an oil firm in your area to see what they offer.

The larger suppliers run planned or automatic delivery services, based on their calculation of the amount of oil you are going to use. Also, they let you pay for the oil in twelve equal payments spread through the year (this saves you having to make a few very large payments).

The maintenance contracts and servicing arrangements are often operated by the oil companies' authorised distributors.

Package deals

Complete, off-the-peg heating systems are available from the gas regions, some electricity boards and oil companies. A fixed price would be quoted, based on the size of the house and the number of rooms, for the supply and installation of a boiler, a given number of radiators, hot water cylinder, cold feed tank, all the necessary pipework and in some cases, some insulation. Anything that you want different from or over and above this basic system, such as sophisticated controls, a more expensive boiler, curved or angled radiators, must be paid for as extras.

You know in advance what the basic cost is going to be, and can choose a system that suits your pocket. But the system you are given is not designed for the details of your particular house. Also, you are restricted in the choice of equipment and the controls for operating the system may be very basic.

There are firms which supply a design for a system based on information from you, the customer. If you agree to go ahead with the design, some of these firms then supply equipment for do-it-yourself installation. The subject was dealt with in *Handyman Which?* in May 1975.

Estimates

You should get an estimate from more than one installer—perhaps three; do not get more because this would be time-wasting for you and the installers. It is easier to get estimates in the spring and early summer when heating contractors are least busy. Ring and ask them to send a representative to discuss what is wanted.

Installing a new central heating system is VAT zero-rated.

An estimate must be in writing and should be comprehensive, preferably including illustrated leaflets, and give details and prices of everything which will be installed, including names of manufacturers, types and models of equipment.

The estimate you accept should include a drawing showing the proposed positions of the boiler, radiators or grilles, pipe runs or air ducts, controls and any other equipment to be supplied. A good installer, when he surveys your house for an estimate, will make detailed notes of all dimensions, structure and construction of the house and existing insulation. He should give a list of temperatures attainable in each room and guarantee that they will be reached when the external temperature is −1°C (30°F).

It should also contain estimates for work carried out by other trades such as building, electrical and insulation work. If any work is necessary that is not covered by the price, the estimate should state this clearly. This means that you should get a full price for the job, and an indication of the length of time that the price holds good for. If there is anything in the estimate that you do not understand, ask what it means.

The estimates are likely to vary considerably in price and some details, even if based on the same requirements. The HVCA *Essential information in estimates for customers* deals with what should be included in an estimate and how it should be presented. If these are followed it will be easier for you, when it comes to comparing estimates, to make sure that any price differences are real and that the estimates relate to exactly the same work.

The dates when work can start and when it will be finished should be given, together with details of when and how you will be expected to pay.

At the time of accepting an estimate it is advisable to agree with the installer that a certain percentage, possibly 5 per cent or 10 per cent of the contract sum, be held back, for say six months, to cover the cost of any defects which may become apparent, particularly if there is no body which backs up the guarantee. Some contractors do not accept such a condition, so you should get it written into the agreement before work starts.

The estimate should provide a specimen of the guarantee offered and tell you what after-sales service is included.

Guarantees

Strictly speaking, in law there should be no need for a guarantee. If you accept an installer's estimate and he agrees to carry out the work based on the estimate, there is a contract between him and you. His part of the contract is to carry out the work properly, your part is to pay him as agreed. If either of you breaks this contract the other can claim for breach of contract. The guarantee does not come into it, but if he offers one, accept it as a useful extra. In practice it is usually much simpler to invoke a guarantee.

There may be other guarantees involved—that of the manufacturer of particular equipment used for the installation: the boiler or the pump, for instance. Your contract with the installer covers everything he installs, boiler and pump and all, so if anything goes wrong with any of them shortly after installation, it is the installer's responsibility to put it right, and if necessary replace a part. This is so in law, whether or not there is a manufacturer's guarantee as well. In practice, if there is a manufacturer's guarantee, it may be simplest to invoke that.

If there is a separate guarantee, check exactly for how long it is valid. Sometimes a guarantee (for a pump, for instance) does not run from the date the pump is installed (as it should) but from the date of its manufacture.

Since the Unfair Contract Terms Act 1977 came into force, it is difficult for any small print conditions in a contract between you and the installer to deprive you of your rights against him. If an exclusion clause in the contract purports to do so, it would only be upheld if it were fair and reasonable in the circumstances and it would be up to the installer to show that his terms were fair, not for you to show that they were not. Moreover, no exclusion clause can affect a claim for compensation if someone is injured (or killed) due to the negligence of the installer.

If after you have agreed an estimate—or even while the work is actually happening—either you or the installer suggest any changes and the other agrees, confirm this (and any extra cost) in writing, rather than rely on the oral agreement. If ever there should be any difficulties, it is much easier to prove what was agreed if there is a written note.

If a dispute does arise between you and the installer it may become necessary to call an independent arbitrator to assess the installation. Some contracts stipulate that such disputes should be referred to the Institute of Arbitrators (75 Cannon Street, London EC4N 5BH). If yours does not, the case can be taken to the Institute if both you and the installer agree. Some professional bodies, such as the Chartered Institution of Building Services, have engineers who are likely to be willing to act in this capacity. Alternatively you can approach a consultancy firm; for instance, the National Heating Consultancy (Gardner House, 188 Albany Street, London NW1 4AP) will appoint a consultant engineer to deal with a technical dispute.

Workmanship
There is likely to be considerable disruption in the house while your new heating system is being installed. There may have to be structural changes to fit the flue; there may need to be some cutting into walls to fit the pipes and install the wiring system for the thermostats.

If you are using a big firm, there will probably be several different people doing different jobs: the main installation work will be carried out by a heating engineer or a plumber, with the help of an assistant. A bricklayer will do the building and chimney work, a joiner deal with any woodwork. Whatever the system, there will be some electrical work to be done by an electrician. With a smaller firm, one or two men are likely to carry out most or all of these jobs.

The British Standards Institution has published a code of practice for *Central heating for domestic premises* (BS 5449: Part 1: 1977, obtainable by post, price £6.40, from 101 Pentonville Road, London N1 9ND) which deals with the work involved in the general planning, designing and installation and includes sections on materials, design considerations, installation work on site. This code represents a standard of good practice which all installers would do well to follow, and HVCA members are committed to follow.

Obviously, you will not be standing over the workmen all the time to watch what they are doing, but there are some points that it is worth looking out for, and if necessary asking for specifically.

All pipework and air ducts should be generally neat and securely bracketed. Make sure that pipes which run under the floor and in out-of-the-way places are not held up with bits of string and nails. Pipe joints should be perfectly clean: any excess solder should have been wiped off. Insulation on pipes and ducts should be neat, particularly at bends.

Where a pipe goes through brickwork or concrete, a pipe sleeve should be fitted around it so that the pipe can expand or contract freely; there should be a metal or plastic cover plate to seal the gap between the pipe and the hole (you might have to pay for this as an extra). The holes through timber or brickwork should be cut with the right size hole saw or drill, and not simply chopped through. All wall fixings should be done with the correct plugs and screws, and not nails.

All bare ironwork, such as the firing door on the boiler and the metal flue pipe, should be painted with heat-resisting paint, otherwise it will rust sooner or later.

Covers should be fitted to all open-top water cisterns to reduce evaporation and to stop dirt getting in. Overflow pipes must be fitted to all water cisterns. They should end outside the house at a point where they can be seen. Overflow pipes larger than 22 mm in diameter (about $\frac{3}{4}$ inch) should be fitted with a wire-mesh guard to prevent birds getting in.

Draincocks should be fitted at all the low points of the pipework where you can get at them easily to enable the entire system to be drained for repairs or shut down in frosty weather.

All rubble, rubbish, and surplus materials should be taken away.

Commissioning and testing
There are two procedures—commissioning and testing—which are not always, but should be, carried out when a new central heating system has been installed. Make sure they are done, preferably in your presence.

Commissioning is the setting up, balancing, and adjustment of the system: adjusting lockshield valves to regulate the flow of water, setting up the oil or gas burner, setting the boiler and room thermostats, control equipment, time switches, duct and grille dampers, fans and any other items of equipment that need attention. All this should be done as part of the actual installing at the very end.

At this stage, any obvious defects should be noted by you and the installer. He should put them right before you settle the account.

Before the installer leaves, see that a full set of cleaning tools for the boiler is left behind, together with an instruction booklet or card for the boiler and other equipment. You should also ask for an air vent key and a lockshield valve key (in case you need to isolate a radiator), and instructions on when and how to use them and how to operate the controls.

Testing is seeing whether all the radiators heat up to the same temperatures and that the controls work properly. It is possible to test the working of the system on midsummer day (provided you do not mind the tropical conditions this will cause). But testing the design should ideally be done when it is cold so that the heating installation can be run under as near design conditions as possible. If each room does not attain the desired temperatures when it is −1°C outside, call back the installer. He has to put right not only workmanship faults but design faults—even if it means that he has to put in new radiators if his calculation of the size of radiators for a particular room was wrong.

Servicing and spare parts

Boilers, warm air units and other items with moving parts cannot go forever unserviced. Regular maintenance can considerably reduce the likelihood of needing expensive replacements. Gas and oil boilers in particular must be regularly serviced even if only from the safety aspect.

Some heating installation firms, oil companies and the nationalised fuel industries offer a servicing agreement whereby the boiler and ancillary equipment are serviced at least once or twice a year for a fee (find out about the cost at the time of getting the estimate).

A *Which?* report on central heating servicing and repairs was published in January 1974.

Most manufacturers of heating equipment are constantly redesigning and altering their products so that items can quickly become out of date or obsolete. Find out about the availability of spare parts before you choose. With little known (and especially imported) equipment, spares may be difficult or impossible to obtain. The larger companies usually carry on making spare parts for some ten years for their major products after these have gone out of production, and merchants may well carry spares on their shelves for longer periods.

It is often worthwhile keeping spares yourself of any small items which may need renewal on a regular basis, so that you can keep the heat going instead of having to wait for spare parts to arrive. Items such as grate bars or oil filters or fan belts for a warm air unit or an igniter for a gas burner do not cost a great deal and will have to be renewed at some stage in the life of the equipment. If you have to buy a replacement for a part that has gone wrong or worn out after a short time, you might as well buy two so that you can keep a spare for the time when this part needs replacing again.

What may appear to be a small defect in your central heating may well involve many hours work including investigations and tests. On the other hand, what may seem to be a major problem can often be put right simply and cheaply. Unless you can find out the cause of the fault and remedy it yourself easily, it is better to call in the installer. Someone who is unfamiliar with the equipment could easily cause damage by tampering with such items as an oil burner or a control box.

lack of heat

If it is considerably colder outside than the temperature for which your system has been designed and you are not as comfortable as you would like to be, you could raise the boiler thermostat from its normal running temperature setting to its maximum setting. At any other time (that is when it is more than $-1°C$ outside), if the boiler will not reach its recommended maximum operating temperature (generally $80°C$), the cause may be a design fault, such as too many radiators for the size of the boiler.

Alternatively, something has gone wrong. The rate at which fuel is being burned may be insufficient to match the designed boiler output. This could be because the boiler has not been correctly set up by the installer, or not properly serviced.

—boiler trouble

The simplest reason why a boiler is not operating is lack of fuel. Do not attempt to dismantle or tamper with burners or controls yourself. They are all made to fail safe if something goes wrong and any alterations could result in fire or explosion.

There are, however, some simple checks you can make. Ensure that fuel is reaching the burner: with an oil boiler, check that the oil tank is not empty and that the vent is clear, because the oil will not flow out of the tank unless air can flow in. Check that the oil filter is not blocked (by removing it and seeing if the oil runs without it), and that there is no air lock in the oil line from the tank to the boiler (by opening the vent on the filter if there is one). If the burner has gone out for some reason, there may be a reset button which has to be pressed. A photoelectric device allows the burner to operate only when there is a flame—this may need to be cleaned. Make sure that the electricity supply to the boiler has not been cut off—check switches and fuses.

With gas, make sure that all gas cocks and valves are in the open position. Check that the pilot light is working. If it keeps going out, the boiler should be seen to by a heating engineer. The cause could be a faulty thermocouple or too low gas pressure resulting in an unstable pilot flame, which could be blown out by draughts within the boiler. (There should be no danger, because the gas supply is automatically shut off. The main burner cannot come on if the pilot light is not established, because of the inherent design of the controls and any small amount of gas that has escaped will vent out through the flue.)

The most likely cause of the fire in a solid fuel boiler burning out is too much draught through the boiler when the dampers are supposed to be closed. If this happens regularly, the installer should check all the seals on the fuelling and ashpit doors and, if necessary, renew them. All dampers should also be checked and the chimney examined for excessive draught. If there is excessive chimney draught, this can be relieved by fitting a draught stabiliser in the base of the chimney to let in air and reduce the draught through the boiler. Conversely, a boiler may go out because there is not enough draught, or even downdraught. You should have the flue and chimney checked to make sure that they are not blocked. Inadequate draught or inadequate air supply can be dangerous because fumes may seep into the house.

With damp or incorrect fuel, the boiler will not burn well. It may go out if the ash or clinker has not been removed or if it has not been correctly banked up at night. This is largely a matter of experience and trial and error with different amounts of fuel.

—controls

If the boiler or warm air unit is operating but the rooms are not at a satisfactory temperature after several hours steady running, some checks should be made to discover the fault. Thermostats occasionally go wrong. If a thermometer is fitted on the boiler, you can check if the boiler thermostat is cutting in and out at the temperature at which it is set.

If your temperatures are controlled by a room thermostat, check that it is set to a high enough temperature level. If a time switch is fitted, see that it is set correctly and is in the 'heat on' period. Check that it shows the correct time—there may have been a power cut while you were out during the day. If you have a mixing valve with a visual indicator, you can make sure that it is in the correct position for maximum heat. Make sure that any thermostatic valves, zone valves or other motorised valves are in the open position; some have a small visual indicator.

—water distribution

Check that the pump is switched on and running. When a pump is working it produces a slight humming sound which you can detect by holding a screwdriver, say, against it (with the handle against your ear, like a stethoscope). If no noise comes out at all, the pump is not working. (If a pump is left unused for weeks or months, it may not start again. Therefore, during the summer, switch it on for five minutes about once a fortnight. To do this, you may have to switch on the boiler control switch to establish the electricity supply, plus any other auxiliary switch for the pump, and make sure that your room thermostat is turned up enough.)

Feel the pipe connections on the flow and return to each radiator. If the flow connection is hot and the return cold or only warm, this indicates poor circulation caused by a faulty pump or wrong size of pump or pipes, or a blockage in a pipe or an incorrectly set lockshield valve. If the flow connection is cold, it is possible that there is no circulation which indicates either a pump fault or an air-lock. If, after venting all the air from the radiator and switching the pump back on, there is still little or no circulation of hot water, the problem obviously lies elsewhere and you must call in the installer.

Check that there is water in the cold feed and expansion cistern. The ball valve may have got stuck. Lack of water can cause air to be drawn into the system.

air-locking

You have to expect air collecting in radiators during and immediately after the initial filling of a system when it is first installed or after it has been drained for any reason. Any further air in a heating system indicates that something is wrong and should be put right as soon as possible. If the pump is too large or incorrectly positioned, air could be constantly drawn into the system. This can also happen if there are any loose valve packings. Gradually, the air would accumulate at the highest points because any air bubbles in water tend to rise.

Radiators are usually above the level of the pipes serving them so that any air starts to collect in the tops of the radiators. If sufficient air collects, it can quite easily prevent the passage of water (even though pumped) past the air bubble. This is known as an air-lock and ought to be traced and got rid of even if it does not seem to affect the heating, because it increases the corrosion rate in the system. Strictly speaking, what is called an air-lock is not always pure air but a mixture of air and gases produced from the interaction between water, air and metal.

Your central heating should be so designed that any air in the system, which is there from the beginning or has subsequently got in, can be vented out either automatically or by hand.

If a radiator is cool at the top and warm at the bottom, air is definitely present. It can be released by opening the air vent a turn or two. First, switch off the circulating pump so that any air which is being carried round by the water will have a chance to separate out and come to the top of the radiator. Open the air vent with the air vent key and hold a jar underneath to catch the water. When it is flowing without air mixed in with it, turn the vent the other way, but do not overtighten. The small diameter thread can be easily stripped and it is no joke having to stand with a finger over the hole until the installer is summoned. Rather than do this, the practical thing to do is to isolate the radiator by turning off the valve each side of it.

Venting a new or refilled system may have to be repeated over a number of days. You should not empty or partially empty the system unless this is really necessary because it introduces more water from which the air and gas mixture must be expelled.

The air vent in a fan convector is usually on the heating pipe inside the casing. Switch off the electricity supply before taking off the cover of the casing to get at the vent.

An air-lock in the pump can be dealt with by using a special key or screwdriver to open—and immediately close—the vent on top of the pump's motor.

An automatic air eliminator can be fitted to replace the ordinary vent valve at the top of a radiator. This will vent any air out of the system automatically but does not cure any underlying fault. Air in a water-filled heating system causes corrosion.

water leaks

If water leaks from a pipe joint, this should be attended to by the installer or a plumber. It may be worth draining the system yourself if there is a delay before the repair man can come.

A leak from the top of a valve can often be tackled by the householder: radiator valves with an adjusting nut underneath the shield should be lightly tightened a sixth of a turn at a time (most nuts are hexagonal) until the leak stops. On a valve with an O-ring seal, tightening does not help—the seal may need to be replaced.

If a leak develops in a one-piece boiler due to corrosion, the whole boiler will probably have to be replaced. Some cast iron boilers are built up from separate sections. If there is a leak from one of these, it is possible to replace just that section, but if it is more than five or six years old it is usually more sensible and cheaper to renew the whole boiler because labour costs are so high and the remaining sections of the boiler are likely to develop leaks within a short time.

Water could leak from a connection between the boiler and the pipework. These are sometimes sealed with a flexible gasket which can deteriorate after a time. If the water looks black, this would not necessarily be detrimental: all water in a heating system becomes discoloured after only a short time in circulation.

Water coming from the feed and expansion cistern's overflow indicates that either the ball valve is faulty or the cistern is too small for the volume of expansion when the system heats up. In the latter case, either a larger cistern should be fitted or the ball valve float in the existing one should be lowered so as to give more room for expansion.

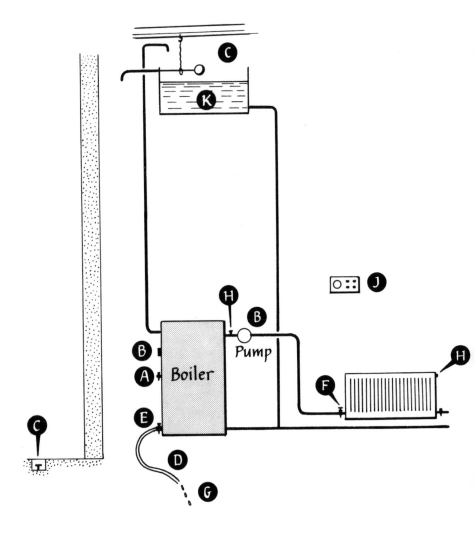

This is how to drain and refill a system:

draining

A   turn off the fuel supply to an oil or gas burner or electric boiler or make sure that a solid fuel fire is out

B   switch off the pump and other electric components, such as the controls

C   tie up the ball valve in the feed and expansion cistern to stop water coming in, or turn off the water supply

D   connect a hose pipe to any draincocks on the boiler or a low point of the pipework, and lead it to a sink or gully

E   open the draincock—you may need to use a spanner or special key

F   open all hand wheel valves

G   allow all water to drain from the system

H   open all air vents on radiators and pipework by two or three turns

refilling

—close all air vents   H

—close the draincocks and remove the hose pipe   ED

—add corrosion inhibitor to the water in the feed and expansion cistern if you had it before   K

—untie the ball valve or turn on the water supply to the feed and expansion cistern   C

—starting at the lower points, open each air vent in turn until water comes through, then shut it   H

—switch on the electricity supply   B

—switch on the pump for five minutes   B

—switch off the pump and vent any more air from the radiators   BH

—switch on the pump again   B

—re-start the boiler   A

—adjust hand wheel valves   F

—check the time switch or programmer   J

—vent again a few hours later and if necessary over the next couple of days   H

A sealed system can be drained by the householder but the refilling and pressurising procedure would need to be carried out by the installer.

noise

Pumps should not be noisy. If there is noise, it could be caused by air in the pump or, more likely, wear or mechanical damage. If the pump is still noisy after venting any air, it should be replaced. (If it keeps failing, this is a sign of a fault in the system, usually that air is being continually drawn in. The sludge which forms through corrosion by air in the system can eventually cause the pump to stop running. The installer would have to find the fault, and flush out the pump.)

Some fans in convectors or warm air units are inherently noisy because they rotate at such a high speed or because of the design of the blades. If the noise gets too much for you, either the whole fan unit has to be replaced or a different pattern of fan fitted. Fan bearings can become noisy, due to wear or lack of lubrication. You should oil the bearings once or twice a year according to the manufacturers' instructions. A fan can also become noisy if it is out of balance due to missing or damaged blades. In this case, it would have to be replaced. If it is out of balance due to the fact that part of an accumulation of dust and fluff has broken away, the fan just needs cleaning.

Gas burners, although not completely silent, are not likely to cause annoyance. Forced draught gas burners and pressure jet oil burners are generally somewhat more noisy, and preferably should not be sited where noise could be a nuisance.

Oil burners should be maintained in a good mechanical condition by regular servicing. The oil burner pump, because of its design, can become very noisy when worn, and should be replaced if this happens. Much wear will be caused if the oil supply is allowed to run out (and the burner tries to run in a dry condition) because the oil pump relies on the fuel for lubrication.

When metal pipes and radiators heat up and cool down, the expansion and contraction of the metal produces a ticking and knocking sound. The problem can be reduced by supporting the pipes with pieces of felt where they rest on the floor joists. There is less noise if non-metallic pipe brackets are used.

A type of resonant sound in the chimney is sometimes produced by pressure jet oil burners due to a combination of boiler and chimney characteristics. The boiler or oil burner manufacturer should be consulted in troublesome cases. The flue may need to be altered, but where this is impossible, there may be no remedy other than installing a different type of oil burner or boiler.

smells and fumes
All lids or doors in a solid fuel boiler or warm air unit should be a good fit to prevent fumes from escaping, and should be replaced if they become warped after prolonged use.

Oil has an unpleasant smell if spilled accidentally or if there is a leak, especially the 35-seconds fuel oil. Any leaks should be attended to by the installer.

There is some smell from any boiler if the chimney is subject to down-draught. In severe cases, expert advice must be sought from the installer. A specially designed chimney terminal or cowl can sometimes overcome minor or intermittent downdraught problems.

poor draught and ventilation
Burning fuel of any sort needs air. Make sure that the appliance can get it. There should be permanent ventilation through air bricks or window ventilators, so see to it that no one blocks a necessary air inlet 'to stop draughts'.

All conventional flue boilers need a good natural draught to operate properly. Generally, the higher and hotter the chimney, the greater will be the draught. The chimney or flue must be large enough to let through the flue gases without too much resistance. If there is a long run of external flue pipe in steel or asbestos, this should be insulated to prevent excessive cooling.

An extractor fan can cause draught problems if it is in the same room as a boiler with a conventional (open) flue. Talk to the installer before putting in an extractor fan.

Chimneys, flues and flueways inside oil and solid fuel boilers should be swept about once a year.

condensation

The flue gases from a gas boiler are relatively cool and contain a lot of water vapour so they will condense on any cool surface. Condensation on chimney breasts usually means that a liner has not been fitted, or, if it has, that it has developed a leak. A builder or heating installer must fit a suitable flue liner to extend to the full height of the chimney.

Condensation in other parts of the house, such as the kitchen or bathroom, can be cured by improving the insulation of walls or windows, increasing the rate of ventilation or raising the room temperature—or a combination of these.

corrosion

Corrosion in radiators is usually caused by air being present in the water. The air may be drawn into the pipework if the pump is too large or in the wrong place or if the system cannot be properly vented.

Rust developing on any part of a heating installation should act as a warning that something is wrong—rust marks on pipes and radiators often indicate that there is a water leak.

Steel radiators are particularly prone to corrosion because the metal is not very thick, and if air is continually taken into the system, they can rust through and become so much scrap metal within a year or two. As metal is corroded away, a black substance from the corrosion process collects at the bottom of radiators and can eventually sludge up the lower parts of the system.

A rusty feed and expansion cistern should be dried out and re-painted with bituminous paint—or, much better, replaced by a plastic one. If you install a new central heating system, have a plastic cistern from the start.

Corrosion is caused not only by the interaction of water with air, but also inside the boiler by water vapour and chemicals from the burning fuel. This is particularly likely to happen if the water returning to the boiler is cool. Therefore, do not set your boiler thermostat below about 60°C (140°F). If the thermostat is set too low, the boiler parts in contact with the flue gases will be too cool to evaporate off the corrosive content of the condensed flue gases.

When two different metals are in contact with each other and water is present, one of the metals will be corroded by electrolytic action. The amount and extent depends on the metals and the temperature of the water. This could happen when copper pipes and a galvanised steel cistern are in contact: the zinc of the cistern will be removed and eventually the underlying steel will corrode through. This may be prevented by fitting a non-metallic coupling, such as nylon, between the metals. Painting the inside of the cistern with bituminous paint also coats the zinc so that chemical reaction will be retarded. When mixed metals are used, corrosion inhibitor should be added to the water in the feed and expansion cistern when it is first filled and whenever it has to be refilled after the system has been drained. Before a corrosion inhibitor is added to the heating water, the local water authority should be consulted. It should never be allowed to mix with the domestic hot water supply, and must therefore not be added in a system where there is any chance of the heating water and domestic water mixing.

scaling
Scale forms when fresh water is continually being brought into the system and heated to above about 60°C (140°F). It should therefore not form where the same water circulates all the time. Scaling should also not be a problem with a domestic hot water cylinder, where the temperature of the tap water is under 60°C. But in very hard water areas there is a possibility of scale forming in the hot water cylinder if the temperature of the tap water (not the heating water) should get excessively hot. Scale which has formed sometimes manifests itself by hissing and knocking noises. It can be softened prior to removal by adding chemical descaler to the water in the cold storage cistern. This is poisonous and must be completely flushed out. It is not a job to be tackled by the householder. In a new system, fitting a water softening unit into the water supply prevents hard scale forming.

It is worth knowing what the possible hazards are with central heating and how to prevent accidents that might occur.

It is a useful precaution with a solid fuel boiler and also with a natural draught pot boiler, to have an extra gravity outlet in the system (possibly a towel rail or radiator) through which water circulates by gravity (and is not pumped), from which excess heat can be dissipated if the pump should fail.

An open vent pipe discharges excess steam or water, to relieve any pressure build-up within the heating system. It leads from either the top of the boiler or from the highest point of the heating flow pipe to above the cistern so that any discharge of water will be into the cistern. The pipe should not end in the open air because under winter conditions it could become blocked with ice. There must be no valve or stopcock anywhere at all between the boiler and the open end of the vent pipe.

safety devices
All oil and gas boilers are supplied with at least a boiler thermostat; some other safety devices may have to be bought separately and fixed by the installer.

One safety device on all gas boilers consists of a thermocouple or flame failure detector which prevents gas entering the combustion chamber if the pilot light is not operating.

Most oil burners are fitted with a flame detector which is linked to the control box. If a flame is not established within a pre-set period, or the flame fails in the course of normal running, the control box shuts the burner off and prevents any more fuel coming into the boiler. The burner can be restarted only by pressing the reset button. This has to be done by hand and so draws attention to the fault. After a burner has stopped firing, or attempting to fire, there should be what is called a purge period of a few minutes so that any unburned vapour can be purged or exhausted from the boiler and chimney. A mechanism in the control box prevents the burner being restarted immediately. This helps to avoid a possible explosion that might be caused by vapour being ignited by any subsequent flame.

Any leak from an oil pipe or oil burner should be attended to straightaway: switch off the supply and call the serviceman. Domestic fuel oil in bulk liquid form and at ordinary temperatures will not burn and can only be induced to do so if it is heated or atomised in some way.

If for any reason there should be a fire anywhere near the boiler, the supply of oil to the boiler must be cut off. There are several kinds of heat sensitive device to do this. Some of these safety devices incorporate a fusible metal link which, when it melts, sets off a mechanism that shuts off the oil supply to the burner. Another type operates from a heat sensitive phial; more rarely used is an electrical type which has a solenoid valve. Whatever the method of operation, the shut-off valve must close automatically when the temperature at the sensing point exceeds 68°C (155°F).

—safety valve
A safety valve should be fitted directly on to, or close to, a boiler. Any undue pressure which builds up inside—perhaps through other safety devices not operating correctly—will cause the valve to open. Any water, steam or air is discharged so that the internal pressure is reduced. A pre-set safety valve is supplied with sealed systems, and every sealed system must be fitted with this type of valve because there is no open vent as an additional safeguard. With a back boiler, if there is not enough space to fit the safety valve to the boiler it can usually be fitted on the flow pipe close by.

A safety valve is best fitted at the time of installation, but it can also be put in later.

gases and fumes
Any trace of smell in the air distributed by a warm air unit must be investigated immediately and attended to. Any toxic fumes would be distributed throughout the house with dire results. Carbon monoxide— the main toxic product of combustion—has no smell: be suspicious of sudden on-going, unexplained headaches and dizziness. The unit should be inspected regularly once a year.

With solid fuel appliances, some fumes inevitably escape when clinkering or removing ash. Ventilation of the surroundings should therefore be adequate to clear these fumes as quickly as possible.

All products of combustion, even if not toxic, can be very unpleasant, and irritating to the skin and eyes. Flue pipes and chimneys must therefore be of suitable materials and large enough to carry away safely the products of combustion to the outside air, and to a height where they cannot get into the windows of bedrooms or other rooms. Chimneys should be extended above the highest point of the building to avoid wind creating any downdraughts. Brick chimneys should be repointed if and when necessary.

Small leakages of gas from pipe joints or similar weak points (which may be fairly difficult to locate) can, over a period of time, build up gas in unventilated spaces.

If you suspect that there is a gas leak, put out cigarettes, do not use any matches or naked flame and open the windows and doors. Make sure it is not simply that a gas tap has been left on accidentally or a pilot light blown out. If you can still smell gas, turn off the main gas supply at the meter, if you can. If you cannot turn off the gas or if the smell persists, telephone the gas region. All gas regions have an emergency service. Its telephone number is listed under 'Gas' in local directories. If the smell goes after the gas has been turned off at the main tap, there is a leak which you must get repaired by a competent person before the gas is turned on again.

flooding

You should inspect your cisterns at least once a year. If they are in poor condition, get a new one without delay. Most cisterns which are now installed are made from plastic or glass fibre material and are therefore immune to corrosion. Make sure, however, that a plastic cistern is properly supported on a flat deck and not just on a couple of joists: when full, an unsupported plastic cistern may bend—or even split.

The other flooding danger with cisterns comes from the ball valve not operating properly so that the cistern overflows. An overflow pipe is therefore normally fitted. It should discharge externally and have a wire mesh guard over the outlet.

Frost can cause leaks in pipes, so all those under ground floors and in the roof space must be insulated. The cold feed and expansion cistern should also be insulated on all sides except the underside, and should have a lid. If the house is unoccupied for more than two or three days in winter the system should, if possible, be programmed to come on for about an hour each day or have a frost thermostat. If the house is to be left unoccupied for a considerable period—more than about three weeks—it would be better to drain the system.

You have a great deal of choice on most aspects of central heating, but there are certain requirements which restrict what you can do. Usually they are concerned with safety and adequate construction.

Some of the requirements are statutory and some should be followed in order to comply with accepted standard practice. Your designer or installer should know of these standards and regulations, and work to them.

Building Regulations

There are separate sets of Building Regulations for England and Wales, Scotland and Northern Ireland which are statutory. Inner London has the London Building Acts and Bylaws.

The Building Regulations cover, amongst other things, the installation of boilers and boilerhouses, chimneys, flues and hearths, and insulation. In some cases, detailed drawings and particulars of the proposed alterations must be submitted to the local authority for approval before any work is started. This is normally done by the installer who is to carry out the work—but the responsibility is yours.

Building Regulations relevant to domestic heating lay down amongst other things the amount of unobstructed permanent ventilation that must be provided to any room that contains a boiler, the qualities and types of materials that may be used for flue pipes, chimneys and hearths, and the way in which these should be constructed in order to reduce fire risk or the risk of fumes getting into the building. The boiler itself must be installed in such a way that excessive heat cannot be transmitted to any surrounding fixtures. The heating installer must ensure that his work complies with all the relevant and up-to-date Regulations.

British Standards

The British Standards Institution issues specifications and codes of practice covering practically all aspects of domestic central heating, including design and manufacture of equipment and the way in which it should be installed. Many, but not all, british-made goods comply with these standards; although the codes of practice are only guides, they should be adhered to by all installers.

The Gas Safety Regulations

These Regulations are designed to protect the general public and individual users of gas and gas appliances, and lay down the responsibilities of installers and users. They are amongst the few sets of regulations which include regulations which you, the consumer, have to follow. For instance, it is your responsibility to see that no appliance is used that you know or suspect to be dangerous. You must ensure that any appliance being used has a proper air supply and a safe flue system. The Regulations also lay down that 'all gas appliances shall be installed by competent persons'. Where a gas installation is put in by direct contract with an installer, without involving the gas region, the local gas service centre will arrange an inspection, at a fee, as a final safety check.

electricity board's requirements

Any electrical work which is carried out in connection with the installation of central heating should comply with the rules of the Institution of Electrical Engineers (known as the IEE Regs). The electricity board should be informed of new wiring in a house and has the authority to inspect any alterations or additions to the wiring. If the board finds dangerous wiring, the supply can be cut off until the fault has been put right.

water board's byelaws

Each regional water authority can make its own byelaws which are usually based on the Model Water Byelaws. These byelaws aim at preventing waste, misuse or contamination of water and include rules about the installation of pipes, storage cisterns and fittings. The water authority has to be informed at least seven days before a new installation using water is put in. You can be fined if the installation does not comply with the byelaws. Generally, pipes and fittings must be protected from frost damage, and fittings must be arranged so that all pipes can be drained if, for example, the premises are to be left unoccupied for any length of time.

local authority

Local authority road parking restrictions may create difficulties for deliveries of solid fuel or oil. Bear this in mind when deciding on the type of fuel to be used.

Planning permission is seldom necessary for an oil storage tank holding less than 3,500 litres (about 770 gallons): check with your local authority. Some, but not all, local authorities have introduced byelaws governing the storage of fuel oil.

Most central heating adds to the rateable value of the property. Whether yours will, depends on the extent to which it is part of the structure of the house. Most removable systems (such as storage heaters) are not rateable. The amount of any rate increase depends on your house and the central heating system, and can add between three per cent and ten per cent to the gross value of the house—but this cannot take effect during the currency of any existing valuation list, that is, not until the next general revaluation.

Improvement grants are given at the discretion of your local authority. You may be able to get a grant towards the cost of installing central heating if it is part of a comprehensive improvement plan for your house—but not if you are installing only central heating or central heating along with other inessential improvements.

An insulation grant of up to £50 or two thirds of the cost (whichever is less) is available to help towards the cost of insulating the roof (if it has no insulation at the moment). Further information, an application form and a booklet *All about loft, tank and pipe insulation* are available free from local council offices.

INDEX

accidents, *see* safety
adjustments, *see* commissioning
advice, 76 *et seq*
Agrément Board, 6
air
– bricks, 8, 97
– eliminator, automatic, 93
– filter, 38, 64
– locks and venting, 44, 46, 87, 89, 91
    et seq, 95, 96
airing cupboard, 60
annular ring, *see* water cylinders, in-
    direct
anthracite, *see* solid fuel
arbitration *see* dispute

back boilers, 22, 23, 28, 31, 32, 34,
    40, 101, *see also* room heaters
background heating, 2, 55, 69, 72
balancing the system, 53, 86, 89
ball valve, 41, 91, 93, 95, 102
boilers, 21 *et seq*, 84, 87, 93, 104
– back, *see* back boilers
– conversions, 32, 33
– dual-fuel, 33
– efficiency, 12, 16, 18, 23, 24, 30
– electric, 16, 34, 72, 95,
– free-standing, 22 *et seq*, 34
– gas, 16, 25, 26, 31 *et seq*, 50, 51,
    87, 90, 95, 96, 98, 100
– jacket, *see* insulation, boiler
– not working, 89, 90
– oil, 16, 24, 25, 26, 29 *et seq*, 50,
    51, 58, 63, 87, 89, 90, 95, 96, 97,
    100
– siting, 24, 25, 26

– solid fuel, 16, 22, 24, 25, 26 *et seq*,
    50, 52, 58, 63, 90, 95, 100
– wall-mounted, 22, 25, 31, 32, 34
bricklayer, 85
British Electrotechnical Approvals
    Board (BEAB), 74
British Gas Corporation, 18, 33, 80
British Standards, 13, 74, 85, 104
British thermal unit, 23
Builders Merchants Federation, 78
Building Centre, 77
Building Regulations, 24, 25, 104
Building Research Advisory Service, 6
building work, 83, 85
bulk supply and storage
– of liquefied petroleum gas, 12
– of oil, 14
bunker for solid fuel, 9, 10, 11
burner, forced draught gas, 32, 96
burners, *see* boilers

catchpit, 13
ceiling heating, 73
centralised oil storage, 14
Chartered Institution of Building
    Services, 76, 85
chemical descaler, 99
chimney, 25, 30, 32, 64, 85, 90, 96,
    97, 98, 102, 104, *see also* flue
circulation
– air in summer, 38, 68
– for domestic hot water, 24, 25, 58 *et
    seq*
– heating water, 41 *et seq*, 91, 100
cisterns, *see* water cisterns
coal, *see* solid fuel
codes of practice, 77, 104 *et seq*

combustion chamber, 22, 65, 100
comfort, 1 *et seq*, 35, 47 *et seq*
commissioning, 86, 87
complaints, 77, 78, 79, 80
condensation, 5, 6, 7, 98
Confederation for the Registration of
  Gas Installers, (CORGI), 80
consultant, 76 *et seq*
consultative councils, *see* electricity
consumers' councils, *see* gas
contents gauge, 14
continuous heating, *see* heat
contract, 12, 79, 80, 81, 83 *et seq*
controls, 21, 47 *et seq*, 87, 90, 91,
  95
– boiler, 47, 50, 51, 52, 57
– compensating system, 49, 56, 57
– convector, 37, 38, 39
– electric heaters, 71 *et seq*
– gas, 32
– modulating, 55, 56, 57
– oil, 13, 14, 30, 31
– radiator, 36
– response to, 44, 46, 49, 57, 73
– solid fuel, 27, 28
– warm air, 64, 68, 69, 70
– zone, 47, 49, 50, 55, 68
see also safety, thermostats, valves
convected heat, 33, 35, 37, 39, 40,
  71, 73
convectors, 21, 35, 37 *et seq*, 41, 45
– fan, 37 *et seq*, 40, 43, 50, 57, 93
– natural, 37
copper coil, *see* water cylinders, indirect
corrosion, 13, 41, 44, 60, 61, 62, 92,
  93, 96, 98, 99
– inhibitor, 95, 99

costs
– running, 4, 9, 16 *et seq*, 24, 47, 67,
  74
– installation, 9, 72, 74, 76, 82, 83,
  84, 87
coupled windows, *see* double glazing
cylinders, *see* water cylinders

dampers, 39, 50, 68, 69, 71
dampness
– in fuel, 9, 10, 90
– in walls, 6
defects, 81, 85, 86, 89 *et seq*
see also workmanship
definitions of central heating, 2
delivery
– of solid fuel, 10, 11
– of oil, 12 *et seq*
design, 3, 23, 47, 49, 67, 76, 77, 82,
  87, 92, 104
– base point, 2, 24
– fault, 87, 89, 90, 96
– temperatures, *see* temperatures
dispute, 84, 85
do-it-yourself installation, 82
domestic hot water, 18, 24, 58 *et
  seq*, 65, 68, 71, 73, 99
– control, 49, 52, 56, 62, 63
– cylinder, *see* water cylinders
Domestic Oil Burning Equipment
  Testing Association (DOBETA), 31
double glazing, 6, 7
draincocks, 86, 95
draining central heating system, 92,
  93 *et seq*, 103, 105, 106
draught, 90, 97
– downdraughts, 3, 7, 40, 68, 69, 90,
  97, 102

— excluder, 7
— stabiliser, 90
drawings, 83, 104
dual-fuel boilers, 33
ducting, 64, 65, 67, 68, 85
— stub, 69
electric heating systems, 16, 64, 65, 71 *et seq*
Electrical Contractors' Association, 79
electrical work, 79, 83, 85, 105
electrician, 85
electricity, 14 *et seq*, 43, 46
— availability, 18, 19
— boards, 16, 73, 78, 79, 82, 105
— consultative councils, 79
— costs, 16 *et seq*, 75
— full price, 14, 16, 34, 71 *et seq*
— off-peak, 14, 16, 34, 64, 71 *et seq*
— standard, *see* full price
— standing charges, 14, 16, 71 *et seq*
— supply, 16, 23, 95
— tariffs, 14, 16
— white meter, 14, 16, 34, 64, 71 *et seq*
electrolytic action, 99
estimates, 78, 82, 83
exclusion clauses *see* contract
expanded plastic foam, 5, 8
expanded polystyrene, 4, 5, 6, 8
expanded polyurethane, 6
expansion vessel,
— sealed system, 41
— self-feed cylinder, 61, 62
explosion, *see* safety
extractor fan, 97

fan, circulating, 16, 28, 32, 35, 37 *et seq*, 50, 64, 68, 71, 72, 96

faults *see* workmanship, defects
feed and expansion cistern, *see* water cisterns
fireplace, 8, 22, 23, 64
fire warning devices, 14, 100 *et seq*
fixed price heating systems, *see* package deals
flame failure devices, 89, 90, 100, 101
flooding, *see* leakage
floor insulation, 7
flow connection/pipe, 20, 44, 45, 46, 53, 54, 91
flow heater, 34, 73
flue, 22, 33, 64, 90, 97, 102, 104, 105
— conventional, 25, 30, 31, 98
— balanced, 25, 26, 30, 31, 40
— fan assisted balanced, 26
foil, aluminium, 4
— backed plasterboard, 5, 6
frost damage, 51, 103, 105
fuels, 9 *et seq*, 95, 100
— changing, 32
— burning rate, 89
— relative costs, 16 *et seq*
— savings, 4 *et seq*, 47 et seq
full central heating, 2
fumes, 12, 25, 90, 97, 101, 102
see also smells

gas, 11, 15
— availability, 11, 18
— boilers, *see* boilers
— consumers' councils, 80
— costs, 16 *et seq*
— maintenance, 11, 32
— meter, 11, 102
— North Sea, 18

– regions, 11, 80, 82, 102, 105
– service pipe, 11
– standing charges, 11, 16
– substitutes for natural (SNG), 18
– tariffs, 11
– warm air heating, 64
see also British Gas Corporation
gas oil, *see* oil
Gas Safety Regulations, 105
glass fibre, 4, 8
grant, improvement, 106
gravity circulation, *see* circulation
gravity-feed boilers, *see* boilers, solid
   fuel
grilles, 64, 67, 68
guarantee, 77, 79, 80, 81, 83 *et seq*

hand-fired boilers, *see* boilers, solid
   fuel
hazards, *see* safety
heat
– boost, 14, 38, 39, 57, 65, 69, 71, 72
– continuous, 2, 3, 51
– distribution, 3, 35, 39, 40, 68, 74
– exchanger, 21, 38, 60, 61, 64, 72
– intermittent, 2, 3
– lack of, 89
– loss, 4 *et seq*, 43, 60, 67
– output, 23, 24, 32, 35 *et seq*, 41, 45,
   50, 71, 72
– requirements, *see* temperatures
– useful, 16
Heating and Ventilating Contractors'
   Association, 77, 83, 85
heating engineer, *see* installer
hinged panels, *see* double glazing

humidity, 3
– control, 70

ignition
– automatic electric, 29, 30, 31
– pilot light, 31, 90, 100
immersion heater, 58, 71, 73, *see also*
   water cylinders
immersion thermostats, *see* thermostats, boiler
installer, 42, 76 *et seq*, 85 *et seq*, 89,
   90, 91, 93, 95, 96, 97, 101, 104 *et
   seq*
– finding him, 76 *et seq*
Institute of Arbitrators, 85
Institute of Domestic Heating and En-
   vironmental Engineers, 77
Institute of Plumbing, 78
Institution of Electrical Engineers
   regulations, 79, 105
instructions, 87
insulation, 4 *et seq*, 35, 74, 83, 104
– boiler, 8, 22
– ceiling, 73
– cisterns, 8, 103
– cylinder, 60, 71
– doors, 7
– ducting, 67, 86
– floors, 7
– flue pipe, 97
– pipes, 8, 60, 86, 103
– roof, 4, 5, 6, 106
– walls, 5, 6, 98
– windows, 6, 7, 98
intermittent heating, *see* heat
isolating valve *see* valves

joiner, 85

kerosene, *see* oil
kilowatt, 23

leakage
— of gas, 90, 102
— of oil, 14, 97, 100, 101
— of water, 41, 42, 93 *et seq*, 102, 103
legal rights, 84, 85
liquefied petroleum gas, 12
local authority, 106
loft insulation, *see* insulation, roof
London Building Acts and Bylaws, 104
lubrication, 96

maintenance, 75, 79, 80 *et seq*, 87, 89, 101
— gas burning equipment, 11, 32
— oil burning equipment, 12, 14, 96
manifold, central, 46
manufacturer's guarantee, *see* guarantee
microbore (minibore) pipe system, 42, 46, 54, 55
mineral wool, 4, 5, 8
Model Water Byelaws, 105

National Association of Plumbing, Heating and Mechanical Services Contractors, 76
National Heating Consultancy, 85
National Inspection Council for Electrical Installation Contracting, 79
natural draught pot boilers, *see* boilers, oil
natural (North Sea) gas, 11
night set-back control, 51, 57
noise, 96, 97, 99
— from boiler, 24, 28, 30, 32, 96
— from fan, 38, 39, 96
— from heat emitters, 38, 39, 40, 96
— from pump, 96
— with warm air heating, 67, 69, 96
— through windows, 7
North Sea oil, 18
nylon pipes, 46

O-ring seal, 93
oil, 12 *et seq*
— availability, 18
— boilers, *see* boilers
— costs, 16 *et seq*
— delivery, 12 *et seq*, 82, 106
— filter, 14, 89
— gas oil (35-second), 12, 29, 97
— kerosene (28-second), 12, 29
— level indicator, *see* contents gauge
— siting tank, 13, 106
— storage, 12 *et seq*, 106
— suppliers, 12, 13, 81, 82, 87
— supply, 12 *et seq*, 18, 89, 96, 101
— types, 12
— viscosity, 12
— warm air heating, 64
one-pipe system, *see* single pipe system
organisations, 76 *et seq*
overfilling alarm system, 13
overflow pipes, 86, 93, 102

package deal heating systems, 82
panel heaters, 73, 74
– plastic, 74
partial central heating, 2
paying
– for electricity, 79
– for gas, 80
– for installation, 83, 84, 86
– for oil, 82
pilot light, *see* ignition
pipe sleeve, 86
pipework, 8, 85, 86, 91, 93, 95, 96, 102, 105
– systems, 41 *et seq*
planning permission, 106
plumber, 85, 93
pressure, 41 *et seq*, 62, 90, 100, 101
– gauge, 42
pressure jet boiler, *see* boilers, oil
prices of fuels, 17 *et seq*
programmer, 49, 57, 62, 95
see also time switch
propane, 12
pump, 16, 21, 29, 32, 43 *et seq*, 50, 53, 57, 58, 62, 72, 84, 91, 92, 93, 95, 98, 100
purge period, 100

quotations, *see* estimates

radiant heat, 33, 35, 40, 73
radiators, 21, 35, 36 *et seq*, 41, 43, 44, 45, 50, 92, 95, 96, 98
– cast iron column, 36
– double panel, 36
– high output, 37
– panel, 36
– shelf, 37

– siting, 37, 43
rate increase, 106
refilling central heating system, 95
regulations, 104 *et seq*
return connection/pipe, 20, 44, 45, 46, 53, 54, 91
road parking restrictions, 106
roof insulation, 4, 5, 6
room heaters, 22, 23, 28, 31, 34, 40
see also back boilers
rust, *see* corrosion

safety, 80, 100 *et seq*
– boiler, 12, 22, 24, 31, 52, 58, 62, 63, 80, 90, 97
– domestic hot water, 62, 80, 99
– electric systems, 71, 73, 74
– flue, 25, 26, 90, 104
– room heater, 40
– warm air units, 65
see also regulations
scaling, 41, 60, 62, 99
sealed units, *see* double glazing
seals, 7, 90, 93
selective heating, 2
servicing, *see* maintenance
Servowarm, 32
single pipe system, 44
skirting heating, 21, 39, 40
sliding panels, *see* double glazing
small bore pipe system, 42, 44 *et seq*
smells, 64, 97, 101, 102
– oil, 12, 97
– gas, 11, 32
smokeless burning
– fuel (Housewarm), 9, 28
– boilers (smoke eaters), *see* boilers, solid fuel

solid fuel, 9, 10, 15
– anthracite, 9, 10
– availability, 18
– boilers, *see* boilers
– coke, 9, 10, 16
– costs, 16 *et seq*
– delivery, 10, 11, 106
– storage, 9, 10
– types, 9
– warm air heating, 64
Solid Fuel Advisory Service, 9, 81
sound-proofing, 7
spare parts, 87, 88
standards, 104 *et seq*
statutory requirements, 104 *et seq*
storage heaters, electric, 14, 16, 71
– water thermal, 72
see also electric heating systems
substitutes for natural gas (SNG), 18
supplementary heating, 2, 40, 50, 72
system
– hot water central heating, 21 *et seq*,
   41, 43 *et seq*
– sealed, 41, 42, 62, 95, 101
– warm air, 64 *et seq*

tap water, *see* domestic hot water
tank, oil storage, 12, 13, 106
temperatures
– controls, *see* controls
– design, 2, 3, 83, 87
– emitter, 36, 37, 40, 41, 44, 45, 74
– for people, 1
– room, 1, 3, 35 *et seq*, 90, 98
temperature detector
– air, 49, 57
– water, 57
testing, 86, 87

therm, 23
thermal cut-out, 71, 74
thermal storage heaters, *see* storage
   heaters
thermocouple, 90, 100
thermometer, 51, 90
thermostats, 49 *et seq*, 56, 72, 90
– boiler, 47, 51, 52, 89, 90, 100
– contact (cylinder), *see* domestic hot
   water control
– limit, 52
– room, 47, 49 *et seq*, 62, 68, 73, 91
35-second oil, *see* oil
throat restrictor, 8
time switch, 16, 32, 38, 40, 47 *et seq*,
   68, 73, 91, 95
see also programmer
towel rail, heated, 59, 100
town gas, 11
see also gas
28-second oil, *see* oil
two-pipe system, 44, 45

underfloor heating, 72
Unfair Contract Terms Act, 84
unit heaters, gas, 40

valves, 53 *et seq*, 58, 62, 90, 91, 93,
   100
– diverting, 62
– double entry, 54
– hand wheel, 53, 95
– isolating, 14, 46, 53, 92
– lockshield, 53, 87, 91
– mixing, 49, 56, 57, 91

– motorised, 48, 50, 55, 57, 62
– null flow (gravity check), 50
– oil drain, 13
– radiator, 42, 45, 47, 93
– safety, 42, 100, 101
– single radiator, 46
– thermostatic control, *see* domestic
   hot water control
– thermostatic radiator, 49, 54, 55
vaporising boilers, *see* boilers, oil
vent pipe, 13, 89, 100
ventilation, 3, 8, 23, 24, 26, 64, 67,
   68, 97, 98, 102, 104, 105
venting, *see* air locks and venting
vermiculite, 4

wallflame boilers, *see* boilers, oil
walls, 98
– cavity insulation, 5
– solid, insulation, 6
warm air heating, 3, 64 *et seq*
– stub duct units, 69
– units, 31, 64, 65, 87, 96, 101
see also ducting, grilles

water
– authority, 99, 105
– distribution, *see* circulation
– jacket, 22, 51, 65
– softener, 99
water cisterns, 5, 86, 102, 103, 105
– feed and expansion (open-topped),
   21, 41, 61, 91, 95, 98, 99
– cold storage cistern, 60, 61
water cylinders, 8, 21, 24, 60 *et seq*,
   99
– automatic self-feed, 61, 62
– controls, *see* domestic hot water
– direct, 58, 60
– indirect, 58, 60, 61
water thermal storage heaters, *see*
   storage heaters
weather compensator, *see* controls,
   compensating system
windows, insulating, 6, 7, 98
wiring, *see* electrical work
woodwork, 85
workmanship, 85, 86

zone control, *see* controls

Illustrations: Neil Hyslop

Extending your house
describes what is involved in having an extension built on to a house or bungalow in England or Wales, explaining what has to be done when and by whom. It deals with planning, professional advice, getting quotations, contracting a builder, how the Building Regulations apply and planning permission. There are many explanatory drawings and a glossary.

The legal side of buying a house
goes through the legal process of buying an owner-occupied house with a registered title in England or Wales (not Scotland). It includes a step-by-step account of a typical house purchase, and also deals with the legal side of selling a house.

An ABC of motoring law
is a guide through the maze of criminal legislation which affects the motorist – in Scotland as well as England and Wales. It describes the offences you could commit, the people and procedures involved, and the penalties you could face. The book is arranged alphabetically, from alcohol to zigzag lines. Clear charts guide you through procedures such as reporting an accident or appearing in court.

Claiming on home, car and holiday insurance
explains the procedure for making a claim on an insurance policy, interpreting the technical jargon and identifying the people and problems you may come across.

How to sue in the county court
goes step by step through what is involved in taking a case to the county court without a solicitor. A woman who sues the shop that sold her a faulty washing machine is used as an example to explain the procedure and rules.

On getting divorced
explains the procedure for getting a divorce in England or Wales, and how, in a straightforward undefended case, it can be done by the postal procedure and without paying for a solicitor. The legal advice scheme and other state help for someone with a low income is described, and there is advice on coping with the home and children in reduced circumstances. Calculations for maintenance and division of property are given, with details of the orders the court may make for financial settlements between the divorcing couple and for arrangements about the children.

Wills and probate
is a book about making a will and about the administration of an estate by executors without the help of a solicitor. One section deals with intestacy and its particular difficulties.

What to do when someone dies
explains the formalities that have to be observed after a death: doctors' certificates, the coroner, registering the death, burial or cremation, the funeral arrangements, national insurance benefits.

Pregnancy month by month
goes in detail through what should happen during pregnancy, mentioning some of the things that could go wrong and what can be done about them, and describing the available welfare services.

Health for old age
sets out in plain language the minor and major physical changes that arise as people grow older, and the treatments available to relieve them. Advice is given about maintaining health and about going to the doctor.

The newborn baby
deals primarily with the first weeks after the baby is born, with information about feeding and development in the following weeks and months. There is advice about when to seek help from midwife, health visitor, clinic doctor to general practitioner.

Earning money at home
is for the person who wants to stay at home while carrying on an occupation which makes some money. It explains what this entails in the way of organising domestic life, family and children, keeping accounts, taking out insurance, coping with tax, costing, dealing with customers, getting supplies. It suggests many activities that could be undertaken, with or without previous experience, and ends with advice about closing down or, more encouragingly, expanding to form a company and employ others.

Cutting your cost of living
suggests ways of spending less money while maintaining your present standard of living. The book not only tells you about shopping sensibly (and lists places where you can buy food cheaply in bulk), but also gives information about growing your own fruit and vegetables. A section on heating suggests ways of cutting down your heating bills and doing jobs yourself, by insulating, by controlling the heat output and by using less hot water.

Living through middle age
faces up to the changes that this stage of life may bring, whether inevitable (in skin, hair, eyes, teeth) or avoidable, such as being over-weight, smoking or drinking too much, insomnia. It discusses the symptoms and treatment of specific disorders that are fairly common in men and women over 40, and for women the effects of the menopause and gynaecological problems. Psychological difficulties for both men and women are discussed, and the possible need for sexual adjustment. Throughout, practical advice is given on overcoming problems.

Having an operation
describes the procedure on admission to hospital: ward routine, hospital personnel, preparation for the operation, anaesthesia, post-operative treatment, convalescence. Basic information is given about some of the more common operations.

Avoiding back trouble
tells you about the spine and what can go wrong with the lower back. It deals with causes of back trouble, specialist examination and treatment and gives hints on general care of the back when sitting, standing, lifting, carrying, doing housework and gardening, and how to avoid becoming a chronic sufferer.

Where to live after retirement
sets out to discuss the problem of suitable accommodation for an older person, with the emphasis on doing something about it while young enough, either by making the present home more suitable, or moving.

This book suggests ways of dealing with the difficulties, and gives sources of help and information for elderly people and their relatives.

Which? way to slim
is the complete guide to losing weight and staying slim. The book separates fact from fallacy, and gives a balanced view of essentials such as suitable weight ranges, target weights, exercise, and the advantages and disadvantages of the different methods of dieting. The book highlights the dangers of being overweight and gives encouraging advice about staying slim.

Dismissal, redundancy and job hunting
discusses the legal implications of being dismissed by an employer and of being made redundant. Recent legislation has affected the position of employed people vis-a-vis employers, their contract of employment and rights on dismissal.

The second part of the book gives advice and suggestions about how to cope with being unemployed and how to set about becoming employed once more.

In preparation: a book about moving house.

Consumer Publications are available from Consumers' Association, Caxton Hill, Hertford and from booksellers.